SUSTAINABLE ASSESSMENT METHOD FOR ENERGY SYSTEMS

Indicators, Criteria and Decision Making Procedure

SUSTAINABLE ASSESSMENT METHOD FOR ENERGY SYSTEMS
Indicators, Criteria and Decision Making Procedure

by

Naim Hamdia Afgan

and

Maria da Graça Carvalho

Instituto Superior Técnico
Portugal

SPRINGER SCIENCE+BUSINESS MEDIA, LLC

Library of Congress Cataloging-in-Publication Data

Afgan, Naim.
Sustainable assessment method for energy systems : indicators, criteria, and decision Making procedure / by Naim Hamdia Afgan and Maria da Graça Carvalho.
p. cm.
Includes bibliographical references and index.
ISBN 978-1-4613-7018-5 ISBN 978-1-4615-4479-1 (eBook)
DOI 10.1007/978-1-4615-4479-1
1.Energy development. 2. Energy Policy. I. Carvalho, M.G. (Maria da Graça), 1955-
II. Title

TJ163.2.A33 2000
333.79'15—dc21 00-056158

Originally published by Kluwer Academic Publishers, New York in 2000
Softcover reprint of the hardcover 1st edition 2000

Printed on acid-free paper.

CONTENTS

PREFACE

The idea for this book was born during our work on The Encyclopaedia of Life Support Systems, which was developed with a group of scientists willing to share their expertise in the promotion of sustainability development on a global scale. It was determined that the complexity of the problem requires a specialised approach and that energy is of paramount importance for the development of the sustainability concept. The multidimensionality of the energy issue has imposed the need for a new approach to the energy strategy concept. In particular, it was recognised that future energy strategy will have to deal with energy as a complex issue encompassing economic, environmental, social, cultural, educational, and resources aspects. This implies that the energy system concept has to be an interactive and proactive, correlating human needs and future developments.

The term sustainability has been "reinvented" in political discourse as a key word to describe quality of life issues, limitation of natural resources and a sense of commitment to our future generations. Other definitions of sustainability have been used in religious, environmental and political contexts in the course of history, each giving specific emphasis to the relevant issues of the time.

The discussion of "limits of growth" focuses on the issue of resource limitations and their potential implications on the future of Western industrial society. While this issue has triggered some positive economic responses in developed parts of the world, it has not yet addressed the growing problems of modern society on a global scale. In an attempt to foster sustainable development, political and normative bodies have focused on the concept of globalisation.

It has become obvious that the inherent complexity of these problems requires researchers and policy makers to enhance and deepen their understanding of the different aspects of present day sustainability. While the economy and the environment are often cited as key issues affecting the quality of life on our planet, emphasis also needs to be placed on the fundamental importance of social aspects in the understanding of sustainable development. As such, attention is now being focused on the links between social processes and other aspects of the modern development strategy.

It is critical to use a reliable methodology for the analysis of a complex system and its behaviour. But before we can do that, it is necessary to define the system, taking into a consideration its complexity. As for any system, this has to be determined by relevant intensive and extensive parameters. This requires a sound knowledge of the internal processes in the system. However, only a system with a known structure can be described with the appropriate number of parameters, thus leading to the description of the internal interaction among the elements of the system. The complexity of the system can not be defined without knowing its structure.

Engineering science has been based on the canonical laws of physics and its derivatives in other branches of science. As such, close links among these fields have been highlighted in a number cases. To a lesser extent, engineering science has been integrated into the social aspects of these issues. The sustainability concept requires a multidisciplinary approach in order to address its full complexity. It is essential to have an understanding of the interactions among the different processes in the system, as a single discipline cannot solely cope with the complexity of these problems.

This book addresses the issue of sustainability by offering a new methodology based on multi-criteria indicators for the evaluation of energy as a system. Since energy plays a critical role in human life, it is imperative to investigate new methods for energy system evaluation. Consequently, the organisation of the book highlights the different indicators for the evaluation of energy systems.

Sustainable Assessment Method for Energy Systems: Indicators, Criteria, and Decision Making Procedure is composed of seven chapters including: Chapter 1- Introduction; Chapter 2 – Sustainability; Chapter 3 – Sustainable Energy Development; Chapter 4- Sustainability and Energy Education; Chapter 5 – Energy System Assessment with Sustainability Indicators: Chapter 6, Sustainability Assessment of Heat Exchanger Design; Chapter 7 – Water and Energy Co-Generation Assessment.

Every human endeavour, including writing a book, requires a supportive environment from those associated with the authors in the process of preparing a book. With the UNESCO Chair on Sustainable Energy Management at the Instituto Superior Tecnico, the Institute was a natural environment for this effort. The authors would like to express their high regard to Prof. B.Berkovski, Director of Engineering and Technology Division of UNESCO and Prof. Dimantino Durão, President of Instituto Superior Tecnico, for their support and vision in the development of this work. The Authors would also like to express high appreciation to Dr. Darwish Al Gobaisi, Editor-in-Chief of the Encyclopaedia for Life Support System, for his encouragement and support. In particular, the authors are indebted to Prof. M.Cumo, University of Rome " La Sapienza", for his suggestions and contribution to formulating the nuclear issues expressed in the book. Also, our great appreciation is devoted to Prof. Mohammad Darvish from Kuwait University for his contribution in presenting material related to the water desalination problem. Prof. N.Hovanov developed some of the decision-making procedures presented in this book and the authors would like to thank him for his co-operation in the preparation of the material for this book.

Every book strongly depends on the support of the people associated with the preparation of the manuscript. The authors would like to express their sincere gratitude to Ms. Rita Maia for her services in editing and typesetting the text and to Mr. Jorge Coelho for his contributions to the graphics and table organisation.

CHAPTER 1

INTRODUCTION

1. INTRODUCTION

Energy resources have always played an important role in the development of the human society [1]. Since the industrial revolution the energy has been a driving force for the modern civilization development. Technological development and consumption of energy, along with the increase in the world population, are interdependent. The Industrial Revolution, especially to the momentum created by the change from reciprocal engines to the great horsepower of steam engines in the late nineteenth century, which brought about a revolution in dynamics - began a drastic increase both in consumption and population of the world [2].

The history of life on the Earth is based on the history of photosynthesis and energy availability [3]. The history of such a planet lies on the capture of the solar energy and its conversion by photosynthesis in plants and phytoplankton as organic molecules of high energy content. The plants convert this energy into other organic compounds and work by biochemical processes. Photosynthesis counteracts entropy increase and degradation since it tends to put disordered material in order. By capturing the solar energy and decreasing the planetary entropy, photosynthesis paves the way for biological evolution.

Boltzman [4], one of the Fathers of modern physical chemistry, wrote, in 1886, that the struggle for life is not a struggle for basic elements or energy, but a struggle for the availability of negative entropy in energy transfer from the hot Sun to the cold Earth. In fact, life on the Earth requires a continuous flux of negative entropy as the result of the solar energy captured by photosynthesis. The Sun is an enormous machine that produces energy by nuclear fusion and offers planet Earth the possibility of receiving large quantities of negative entropy. Every year the Sun sends 5.6×10^{24} joules of energy to the Earth and produces 2×10^{11} tons of organic material by photosyntesis. This is equivalent to 3×10^{21} joules/ year. Through the billions of years since the creation of the planet Earth this process has led to the accumulation of an enormous energy in form of different hydrocarbons. Most of the fossil fuels belong to the type of material where molecular binding is due to Van der Waals potential between every two molecules of the same material. Mankind's energy resources rely heavily on the chemical energy stored in the fossil fuel. Table 1 shows assessed energy resources [5].

Energy and matter constitute the earth's natural capital that is essential for human activities such as industry, amenities and services in our natural capital as the inhabitants of the planet earth may be classified as:

- Solar capital (provides 99% of the energy used on the Earth)
- Earth capital (life support resources and processes including human resources)

These, and other, natural resources and processes comprise what has become known as 'natural capital' and it is this natural capital that many suggest is being rapidly degraded at this time. Many also suggest that contemporary economic theory does not appreciate the significance of natural capital in techno-economic production.

2. ENERGY RESOURCES

All natural resources are, in theory, renewable but over widely different time scales. If the time period for renewal is small, they are said to be renewable. If the renewal takes place over a somewhat longer period of time that falls within the time frame of our lives, they are said to be potentially renewable. Since renewal of certain natural resources is only possible due to geological processes which take place on such a long time scale that for all our practical purposes, we should regard them as non-renewable. Our use of natural material resources is associated with no loss of matter as such. Basically all earth matter remains with the earth but in a form in which it can not be used easily. The quality or useful part of a given amount of energy is degraded invariably due to use and we say that entropy is increased.

There are number of sources for the date on energy resources and reserves. There are two factors effecting the reliability of date. Time the data are produced is one of the substantial factor which reflects the amount of capital invested in the investigation of new resources. The second factor which may be linked to the time is the technology available for the discovery of new resources. With these two constrains, the data presented in Table 1, are taken from WEC sources.

Table 1. Fossil Fuel and Uranium Reserves at the end of 1993[4]

	Proven Reserves[1] Gtoe[5]	Resources[2] Gtoe	Resource Base[3] Gtoe
Oil	343	477	820
Natural Gas	141	279	420
Coal & Lignite	606	2,794	3,400
Total Fossil Fuel	1,090	3,550	4,640
Uranium	In Thermal Reactors	In fast Reactors	

	Proven Reserves[1] Gtoe[5]	Resources[2] Gtoe	Resource Base[3] Gtoe
	3,390		15,540

1 Proven Reserves are those that can be produced with existing technologies under present market conditions.
2 Resources are those which with technical progress could become economically attractive to produce.
3 Resource Base is Proven Reserves plus Resources.
4 Source:'Global Energy Perspectives to 2050 and Beyond', WEC/IIASA, 1995.
5 Gtoe: Giga tonnes of oil equivalent

Use of the energy resources is subject to the technologies for their transformation. In this respect Table 2 gives schematic presentation of the technologies needed to convert the energy resources in the form adapted for its use.

Table 2. Schematic presentation of the energy resource technologies.

Resources	**Coal**	**Gas**	**Oil**	**Nuclear**	**Waste**	**Solar**	
Extraction	Coal mine	Gas well	Oil well	Nuclear processing	Agro-forestry	Collector	
Primary sources	Coal	Gas	Oil	Nuclear fuel	Bio-mass	Heat	Electricity
Conversion Technology	Power plant	Gas distribution		Refinery	Methanol Plant		
Distribution Technology	Electric grid	Gas grid		Oil transport system			
Final Energy	Electricity	Gas		Kerosene			
End use technology	Engine	Furnace	Light	Oven	Automobile		
Energy service	Power	Space heating	Illumination	Cooking	Transport	Space heating	Illumination

Adapted from N.Nakicenovic at al [5].

3. ENERGY CONSUMPTION - PAST, PRESENT AND FUTURE

The abundant energy resources at the early days of the industrial development of the modern society have imposed the development strategy of our civilization to be based on the anticipated thinking that energy resources are unlimited and there is no other limitations which might affect human welfare development. It has been recognized that the pattern of the energy resource use has been strongly dependent on the technology development. In this respect it is instructive to observe [6,7] the change in the consumption of different resources through the history of energy consumption.

Worldwide use of primary energy sources since 1850 is shown in Figure 1 [6]. F is the fraction of the market taken by each primary-energy source at a given time. It could be noticed that two factors are affecting the energy pattern in the history.

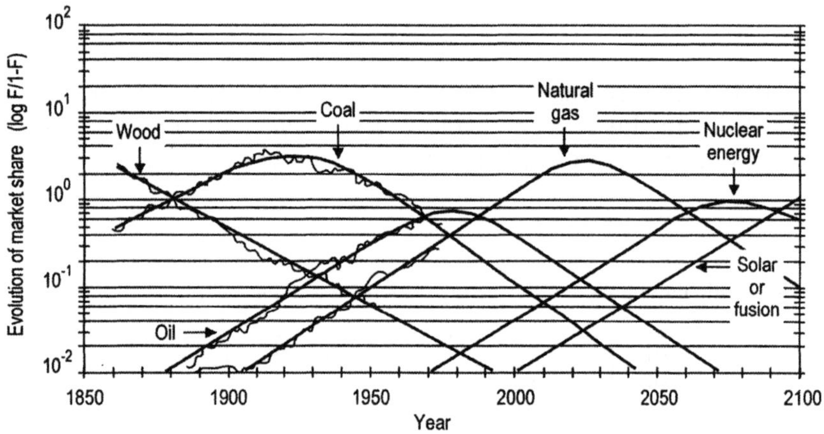

Figure 1. Primary energy sources.

The first is related to the technology development and, the second, to availability of the respective energy resources. Obviously, this pattern of energy source use is developed under constraint immanent to the total level of energy resources consumption and reflects the existing social structure both in numbers and diversity [8,9,10]. The world energy consumption is shown in Figure 2. [11].

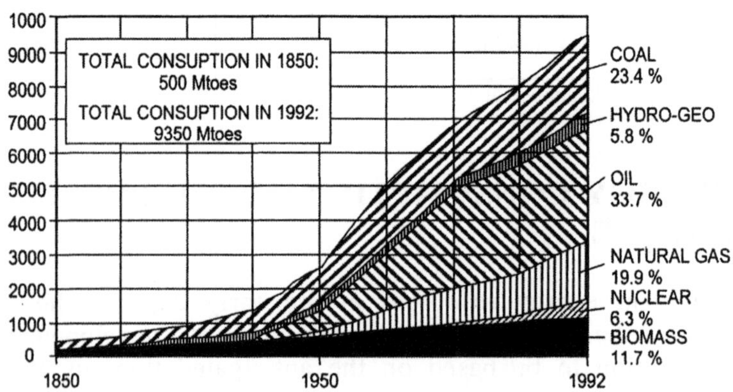

Figure 2. World energy consumption.

Looking at the present energy sources consumption pattern, it can be noticed that oil is a major contender, supplying about 40% of energy. Next, coal supply is around 30%, natural gas 20% and nuclear energy 6.5%. This means that current fossil fuel supply is 90 % of the present energy use. In the last several decades our civilization has witnessed changes which are questioning our long-term prospect. Fossil fuel, non-recyclable is an exhaustible natural resource that will be no more available one day. In this respect it is of common interest to learn how long fossil fuel resources will be available, as they are the main source of energy for our civilization. This question has attracted the attention of a number of distinguished authorities, trying to forecast the energy future of our planet. The Report of the Club of Rome " Limits to Growth", published in 1972 [12], was among the first ones which pointed to the finite nature of fossil fuel. After the first and second energy crisis the community at large has become aware of the possible the physical exhaustion of fossil fuels. The amount of fuel available is dependent on the cost involved. For oil it was estimated that proved amount of reserves has, over past twenty years, leveled off at 2.2 trillion of barrels produced under $ 20 per barrel. Over the last 150 years we have already used up one-third of that amount, or about 700 billion of barrels which leaves only a remaining of 1.5 trillion of barrels. If compared with the present consumption, it means that oil is available only for the next 40 years. Figure 3 shows the ratio of the discovered resources to the yearly consumption for the fossil fuels [11].

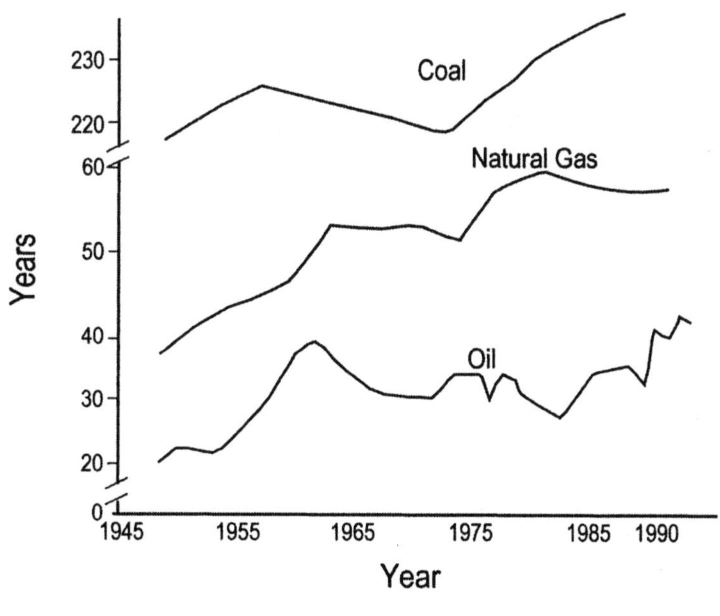

Figure 3. The ratio of the discovered resources to the yearly consumption for the fossil fuels.

From this figure it can be noticed that coal is available for the next 250 years and gas for the next 50 years. Also, it is evident that as much as the fuel consumption is increasing, new technologies aimed to the discovery of new resources are becoming available, leading to a slow increase of the time period for the exhausting of the available energy sources.

It is known that the energy consumption is dependent on two main parameters. Namely, the amount of energy consumed per capita and the growth of population. It has been proved that there is a strong correlation between the Gross Domestic Product and Energy consumption per capita. Figure 4 shows the economic growth and energy consumption for a number of countries, in 1990 [12].

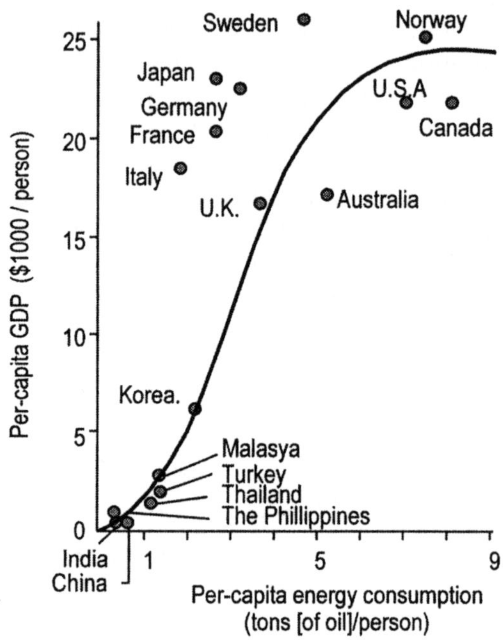

Figure 4. Economic growth and energy consumption for a number of countries, in 1990.

There is a number of scenarios which are used for the forecast of the world economic development. With the assumption that the recent trend in the economic development will be conserved in the next 50 years and considering the demographic forecast in the increase of human population, as shown in Figure 5 [13] the future energy consumption could be calculated, as shown in Figure 6 [14].

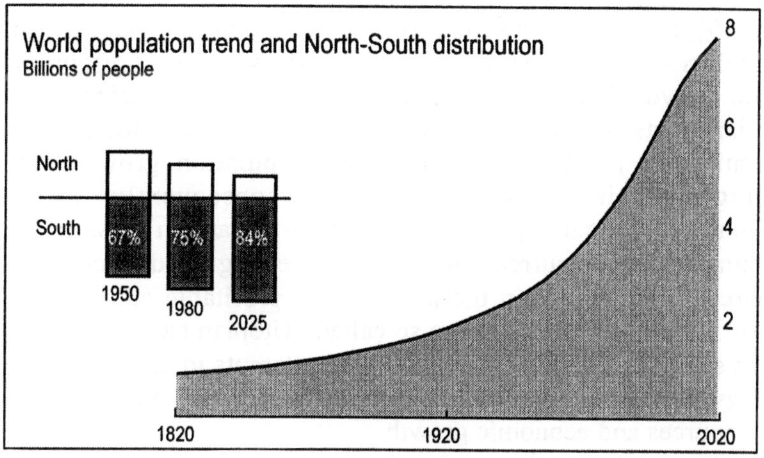

Figure 5. Demographic forecast in the increase of human population.

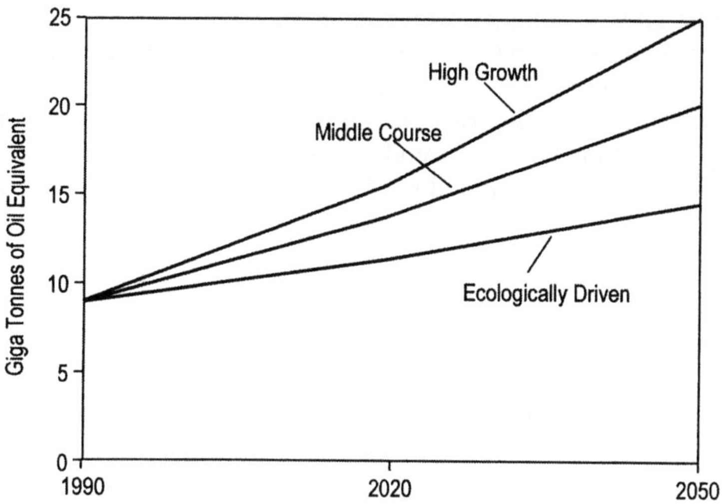

Figure 6. Future energy consumption.

Compared with the available resources it is easily foreseen that the depletion of the energy resources is an immanent process which our civilization will face in the near future. Nevertheless, whatever is the accuracy of our prediction methods and models, it is obvious that any inaccuracy in our calculation may affect only the time scale but not the essential understanding that the energy resources depletion process has begun and requires the human action before adverse effects may irreversibly enforce.

Natural resources scarcity and economic growth are in fundamental opposition to each other. The study of the contemporary and historical beliefs showed [15], that: (1) natural resources are economically scarce, and become increasingly so with the passage of time; (2) the scarcity of resources opposes economic growth. There are two basic versions of this doctrine. The first, the Malthusian, rests on the assumption that there are absolutely limits; once these limits are reached the continuing population growth requires an increasing intensity of cultivation and, consequently, brings about diminishing returns per capita. The second, or Ricardian version, viewed the diminishing returns as current phenomena reflecting the decline in the quality of resources brought within the margin of a profitable cultivation. Besides these two models, there is also the so called "Utopian case' where there is no resources scarcity. There have been several attempts to apply these models to the energy resources in order to define the correlation between specific energy resources and economic growth.

The substantial questions related to the scarcity, its measurement and growth are: (1) whether the scarcity of energy resources has been and/or will continue to be mitigated and (2) whether the scarcity has "de facto" impacted the economic growth. An analysis based on the relative energy prices and unit costs has been applied to natural gas, bitumen coal, anthracite coal and crude oil. The USA analysis in this respect can serve as the indication for the future trend in a world scale. Using two measures of scarcity - unit cost and relative resource price change in the trend of resource scarcity for natural gas, bitumen coal, anthracite coal and crude oil, over three decades are shown in Figure 7 [16].

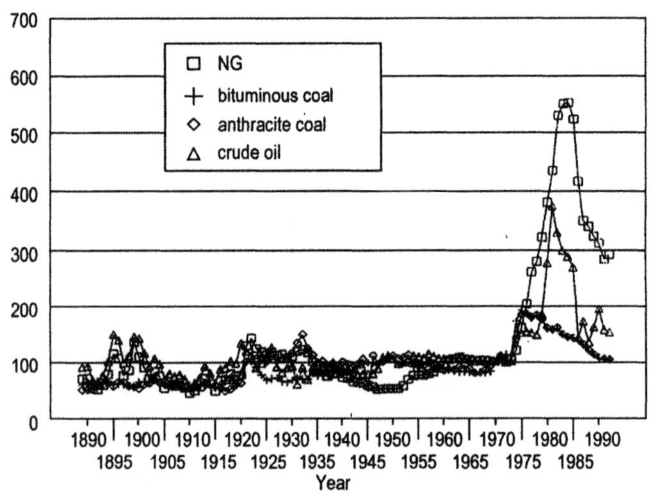

Figure 7. Resource scarcity for natural gas, bitumen coal, anthracite coal and crude oil.

It can be noticed that each one of the energy resources has become significantly more scarce during the decade of the 1970s. The situation reversed itself during the 1980s. The change, that took place, has implications for the future economic growth to the extent resources scarcity and economic growth are interrelated, even if it was not proved that short term energy resources scarcity fluctuation has substantial implication on the long term economic growth. It has become obvious the need for an active involvement in allocating scarce, non-renewable energy resources and its potential effect on the economic growth.

4. ENVIRONMENT CAPACITY CONSUMPTION - PAST, PRESENT AND FUTURE

Primary energy resources use is a major source of emissions [17,18,19,20]. Since fossil fuels have demonstrated their economic superiority, more than 88 % of primary energy in the world in recent years has been generated from fossil fuels. However, the exhaust gases from combusted fuels have accumulated to an extent where a serious damage is being done to the world global environment. CO_2 is the mostly debated one of the greenhouse gases. Its global warming potential is compared to the other greenhouse gases is lower but its emission are higher and are directly related to a daily need of humankind, namely energy supply. It is therefore very important that its sources and sinks are well understood and investigated. The balance and the assessment of implications of changes of CO_2 is of great importance. The quantification of how much is taken up by each CO_2 sink is complicated and cause to numerous re-evaluations. The recent estimations suggested that about 2 Gt C/yr is taken up by ocean, about 3.2 Gt C/yr by atmosphere and about 1.4 Gt C/yr is assimilated up by terrestrial ecosystem. The accumulated amount of CO_2 in atmosphere is estimated at about 2.75×10^{12} t. The global warming trend from 1900 - 1990 is shown in Figure 8 [21].

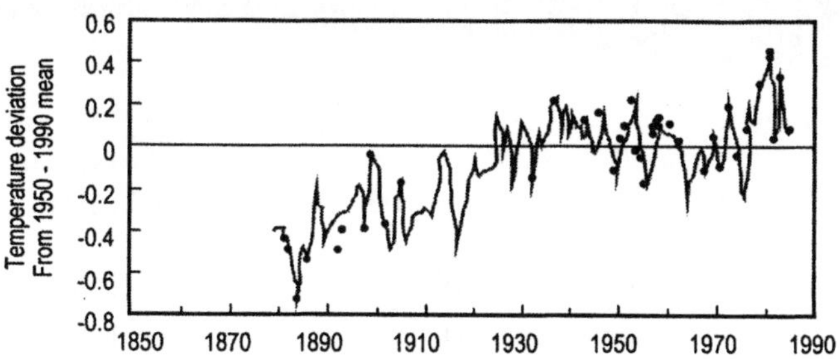

Figure 8. Global warming trend from 1900 – 1990.

The future trend of the carbon dioxide concentration in the atmosphere can be seen from the Figure 9 [21].

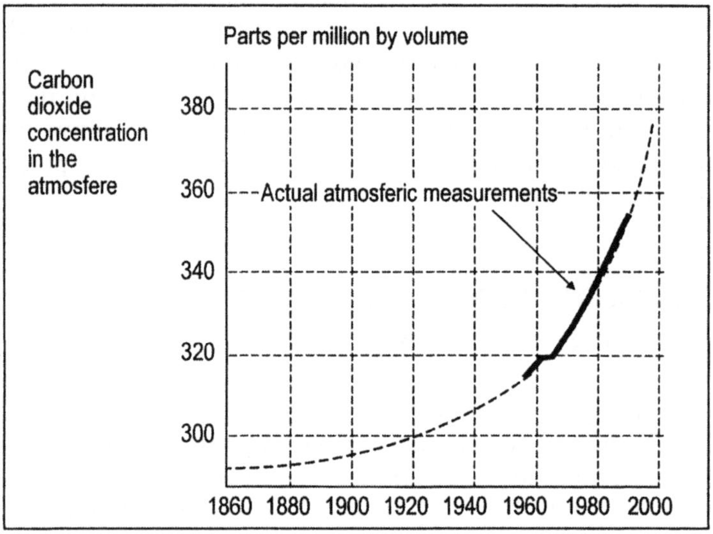

Figure 9. Carbon dioxide concentration in the atmosphere.

It is rather obvious that the further increase of the CO_2 will lead to disastrous effects to the environment. Also, the emission of SO_2, NOx and suspended particulate matters will substantially contribute to exasperate the effect on the environment.

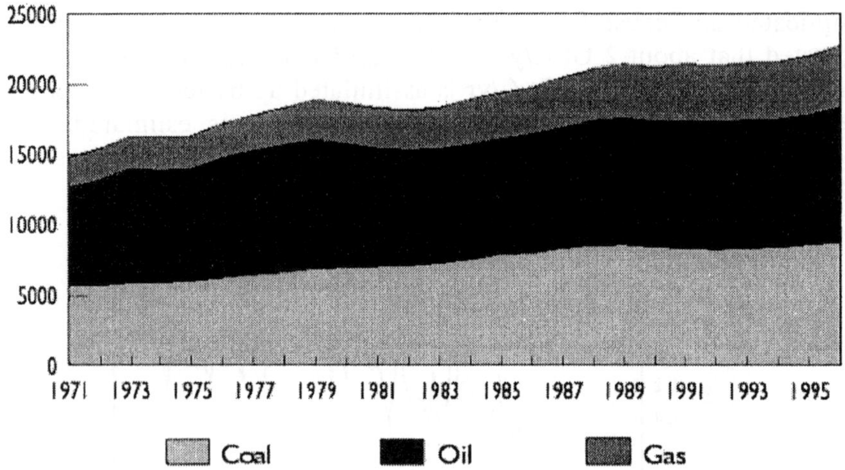

Figure 10. Change in the CO_2 production for the last 20 years.

Figure 10 presents the change in the CO_2 production for the last 20 years [22]. It can be noticed that coal combustion has major contribution to CO_2 production.

In a world scale, coal will continue to be a major source of fuel for the electric power generation. Many developing countries, such as China and India, will continue to use inexpensive, abundant, indigenous coal to meet growing domestic needs [23,24,25,26]. This trend greatly increases the use of coal worldwide as economy in the other developing countries, continues to expand. In this respect the major long-term environmental concern about coal use has changed from acid rain to greenhouse gas emissions - primarily carbon dioxide from combustion. It is expected that coal will continue to dominate China's energy picture in the future. The share of coal, in primary energy consumption is forecast to be no less than 70 % during the period 1995 - 2010. In 1993 China has produced a total of 1.114 billion tons of coal, in 2000 it is planned 1.5 trillion and in 2010 it will be 2.0. Since China is the third largest energy producer in world, after USA and Russia its contribution to the global accumulation of the CO_2 will be substantial if the respective mitigation strategies will not be adopted. The example of China is instructive in the assessment of the future development of developing countries and their need for accelerated economic development.

Carbon intensity of finale energy expressed in tones of carbon per ton of oil equivalent energy is parameter which give us evidence of the contribution of different countries. Comparing the carbon intensity of finale energy for the United State, Japan, and France in recent decades shows that the three industrialized countries have spared about 10 percent, but at the same period of time China and India have increased its contribution of CO_2 to the global production by more then 20 %. Figure 11 shows contribution of different regions to the total CO_2 production.

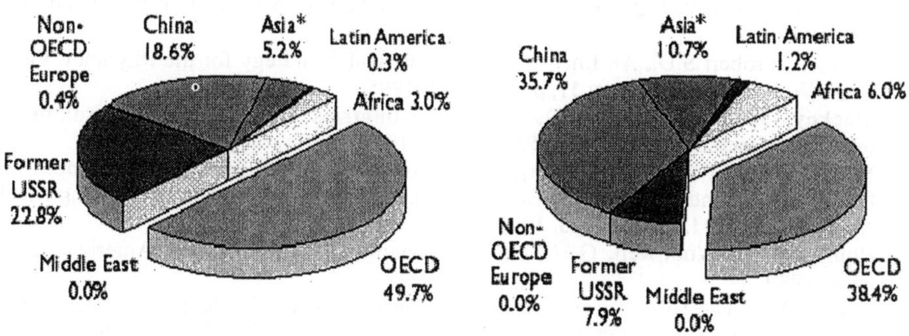

Figure 11. Contribution of different regions to the total CO_2 production.

5. REFERENCES

1. Marchetti C., From Principles to World Government, International Institute for Applied System Analysis, 1995.
2. Marchetti C., Check on the Earth - Capacity for Man, Energy,Vol. 4, pp.1107-1117, 1979.
3. Ohta T., Energy Technology, Elsevier Science, 1994.
4. Botzman, L., Vorlesung zur Gas Theorie, Vol.1, Leipzig, 1896.
5. Nakicenovic, H., Gruber, A., McDonald, A., Global Energy Perspective, Cambridge University Press, Cambridge, UK, 1998.
6. Master, C.D., World resources of Crude Oil, Natural Gas, Bitumens and Shale Oil. Topic 25, World Petroleum Congress Publ, Houston, Ref. Table 1, 1987.
7. Marchetti, C., The " Historical Instant " of Fossil Fuel, Symptom of Sick World, Int. Journal Hydrogen Energy, 16, pp.563-575, 1991.
8. Marchetti, C., Society as Learning System,Technol. Forecast. Soc. Changes, 18, pp. 267-282, 1980.
9. Arnold, M.St.J., Kersall, G.J., Nelson, D.M., Clean Efficient Electric Generation for the Next Century: British Coal Toping Cycle, Combustion Technology for Clean Environment, Ed: M.G.Carvalho, W.A.Fineland, F.C. Lockwood, Ch. Papaloupolos, 1995.
10. Mazzuracchio, P, Raggi, A, Barbiri, G., New Method for Assessment the Global Quality of Energy System, Applied Energy 53, pp.315-324, 1996.
11. Noel, D., A Recommendation of Effect of Energy Scarcity on Economic Growth, Energy, Vol. 2, pp. 1-12, 1995.
12. Medows, D.H., Meadows, D.L,, Randers, J., Behrens III, W.W., The Limits of Growth, Universe Book, New York, 1972.
13. Farinelli, U., Alternative Energy Sources for the Third World: Perspective, Barriers, Opportunity, Pontifical Academy of Science, Plenary Session, Oct.25-29, 1994.
14. Keatiny, M., Agenda for Change, Center for Our Common Future, 1993.
15. WEC Message for 1997, Briefing Notes.
16. Barnett, H.J., Morse, Ch., Scarcity and Growth, Resources for Future, Inc., 1963.
17. Noel, S., A Reconsideration of Effects of Energy Scarcity on Economic Growth, Energy, Vol. 20, No. 1, pp. 1-12, 1995.
18. Mackey, R.M., Probert S.D., National Policy for Achieving Thrift, Environmental Protection, Improved Quality of Life and Sustainability, Applied Energy 51, pp. 243-367, 1995.
19. Price, T., Probert S.D., An Energy and Environmental Strategy for the Rhymney Value, South Walls, Applied Energy 51, pp. 139-195, 1995.
20. Mackey, R.M., Probert, S.D, NAFTA Countries Energy and Environmental Interdependence, Applied Energy 52, pp. 1-33, 1995.
21. Mackey, R.M., Probert, S.D, Energy and Environmental Policies of the Developed and Developing Countries within the Evolving Oceania and South-East Asian Trading Block, Applied Energy 51, pp. 369-406, 1995.
22. Hought, R.A., Woodwell, G.M., Global Climatic Change, Scientific American, April issue, 1989.
23. Nunes, P.S., The role of Technologies Facing the Kyoto Challenge, 5th International Conference on Technologies and Combustion for a Clean Air, July 12-15, Lisbon, 1999.
24. Wu, K., Li, B., Energy Development in China, Energy Policy, Vol. 23, No. 2, pp. 167-178, 1995.
25. Painuly, J.P., Rao, N., Parickh, J.,A., Rural Energy-Agriculture Interaction Model Applied to Karnatak State, Energy, Vol. 20, No. 3, pp. 219-233, 1995.

26. Hollander, J.M., Schnaider, T.R., Energy Efficiency: Issues for the Decades, Energy, Vol. 24, No. 4, pp. 273-287, 1998.

CHAPTER 2

SUSTAINABILITY

1. INTRODUCTION

As it has been shown, the energy resources are the bricks for building our civilization [1]. Their polyvalent use has offered a service to human society, leading to the welfare commodity for a decent level of human life. Sadly, however, production and consumption of energy are going hand in hand less than welcome side effects. This is the reason why society has recognized the importance of intelligent energy use with a sensibility that the required energy services be provided as clean and efficient as possible. Crucial importance is added due to the rapid growing world population and the need for accelerated economic development of developing countries. This is the reason why energy takes a centre stage in the debate surrounding one of today's main dilemmas: how to combine economic development with a habitable environment in a world that is undergoing rapid changes as a result of population growth and economic development of the developing part of the world.

Lately, "sustainability" has become a popular buzzword in the discussion of the resources use and environment policy. The word sustainability has root in the Latin – sustinere-, meaning "to hold up" by world inhabitants, present and future ones. Before any further discussion of the subject, it is necessary to define and properly assess the term we are going to use. It should be emphasized that the definition is needed in order to clarify the concepts. So, what is sustainability?

2. PRESENT DEFINITION

Among the terms most often adopted there are the following:

a.) for the World Commission on Environment and Development (Brundtland Commission) [2]:
"development that meets the needs of the present without compromising the ability of future generations to meet their own needs"

b.) for the Agenda 21, Chapter 35 [3]:

"the development requires taking long-term perspectives, integrating local and regional effects of global change into the development process, and using the best scientific and traditional knowledge available"

c.) for the Council of Academies of Engineering and Technological Sciences [4]:
"It means the balancing of economic, social environmental and technological considerations, as well as the incorporation of a set of ethic values"

d.) for the Earth Chapter [5]:
"The protection of the environment is essential for human well-being and the enjoyment of fundamental rights, and as such requires the exercise of corresponding fundamental duties"

e.) Thomas Jefferson, Sept. 6 1789:
"Then I say the earth belongs to each generation during its course, fully and in its right no generation can contract debts greater than may be paid during the course of its existence" [6]

All five definitions stand for the emphasis of specific aspects of sustainability. Definition a.) and e.) imply that each generation must bequeath enough natural capital to permit future generations to satisfy their needs. Even if there is some ambiguity in this definition, it is meant that we should leave our descendants the ability to survive well and meet their own needs. Also, there is no specification in which form resources are to be left and how much is needed for the future generations, because it is difficult to anticipate the future scenarios.

Definitions b.) and c.) are more politic plies for the actions to be taken at global, regional and local levels in order to stimulate the United Nation Organization, Governments and Local Authorities to plan development programs in accordance with the scientific and technological knowledge. In particular, it should be noticed in definition c.) the ethic aspect of the future development actions to be taken to meet sustainable development.

Definition d.) is based on the religious believes playing the responsibility and duties towards the nature and the Earth. In this respect it is of interest to enlighten that the Old Testimony, in which the story of creation is told is a fundamental basis for Hebrew and Christian doctrines of the environment. In the world of Islam, nature is the basis for human consciousness. According to the Koran, while humankind is God's vice-regent on Earth, God is the Creator and Owner of nature. But human beings are his trusted administrators, they ought to follow God's instructions, that is, acquiesce to the authority of the Prophet and to the Koran regarding nature and natural resources.

3. SCIENCES AND SUSTAINABILITY

The Club in Rome in the treaty "Limits of Growth" [7] has shown that our human society is exposed to the limits of different resources. Even, it was proved that numerical values presented are not very accurate, it opens new issues and has focused our attention to the global problems on our planet. Number of scholars have joined this movement and have enlighten different aspect of global problems, showing that besides the resources limits there are also environment, social and economic limitations of interest to be better understood.

The Brundtland Commission on Environment and Development in its report "Our Common Future", [8] has come out with the first modern definition of sustainability.

The United Nation Organization Conference on Environment and Development was held in Rio de Janeiro, attended by 110 head of the states has adapted document Agenda 21 [3], with the main principles which are in compliance with the sustainable development. In the Agenda of Science for Environment and Development into 21st Century the following principles are defined:

a.) Strengthening the scientific bases for sustainable management;
b.) Enhancing scientific understanding;
c.) Improving long-term scientific assessment;
d.) Building up scientific capacity and capability.

It has been emphasized the substantial role scientific and technological advancement in the improvement of quality of life on our planet. Even there has been shown a high political will to address the problem and contribute to its solution, the period after the Conference has not been used to the qualitative improvement in the world. Further deterioration of the "natural capital", irreversible changes in the environment, degradation of social structure and further increase in differénces between the North and South have lead political leaders to initiate new actions in order to prevent further degradation of global environment.

After several international conferences, the Kyoto Conference on Global Climate Change [9] has addressed specifically the ever rising problem of the environment pollution. With the adapted Resolution, it was agreed to set limits, which are to reflect the need for the conceivable actions on global, regional, state and community level. It was proved to be active movement with the political conciseness with the irreversible changes in global environment. Numeral values of limit of CO_2 concentration are an indicator to serve as a scale for assessment of the successes of political action. This shows a good example for the control of achievement obtained by the specific political plies. The CO_2 concentration is single parameter, which is easy to measure, and its function as the environment indicator is of cardinal importance but far from being sufficient for the sustainability assessment.

3.1 Modern Concept of Sustainability

The sustainability has been reinvented as key word to describe for a political discourse concerning quality of life issues, limitation of natural resources and a sense of the commitment to future generation [10]. It is known that many individuals reflecting different issues, which they had in mind, have defined the sustainability [11]. Even sometimes controversial these definitions have served as the discourses in religious, environmental and political plies as a set of normative and strategic issues of the respective doctrine. In this respect there have been changes in the course of history which has given specific emphasize to the specific issues relevant for the respective period of the history.

Our ever growing interest have lead us to the redefinition of the main discourse in our development strategy in order to adapt our future to the irreversible changes which are immanent to our civilization. As an example we can see the sustainability development closely linked to the rejection of the "development " and mere economic modernization as a new approach for the management of social structure change. On the one hand, "development" and "modernization" have attracted growing criticism as a mean of domination from those not belonging to the developed part of the world. On the other hand, "development" as a generalization of the historical development of Western world has proved inappropriate to meet the most urgent need of present global transformation process.

The discussion of the "Limits of growth" was focused to the single issue reflecting the resource limits, which may effect the Western industrial society future if the same path will be followed on the global scale. Even, it has triggered some positive response in the total economy of the developed part of the world; it has not given answers to the growing problems in the modern society. At that time it was launched "globalization" as a key word for the political and normative consensus to foster sustainable development. At the same time, due to uneven economic and political segmentation and fragmentation, poverty and marginalization are a threatening growing section of the population.

The economic vision of sustainability related to the energy resources is based on two approaches. First is based on the sustainability between the natural capital and man-made capital. The second approach is based on considering man-made capital and natural capital as compliments. It is seen that the first approach is not acceptable, since it does not take into account the entropy law with regards to the scarcity of resources and does not consider the finitude of terrestrial deposits [11]. This vision of sustainability is reflecting only resources aspect of sustainability and neglecting the environment and social affect of the notion of sustainability. Obviously, simplification in this definition of sustainability is not taking complexity of the global system as relates to the energy resources. As the example, to what extend the social aspect of sustainability is effecting the energy resource use can be seen through the public media affect to the energy consumption in the cases of energy crises.

It has become evident that the complexity of the problems requires the need to enhance and deepen the understanding of the implication of different aspects of the sustainability. Besides the resources, economy and environment as the key issues of the problems effecting quality of life on our planet it has been recognized that the social aspect is of fundamental importance in understanding the sustainable development. The attention was focused on the links between social processes and other aspects of the modern development strategy.

It is immanent to the complex system to use respective methodology in the analysis of its behavior. But, before we can do that, it is necessary to define the system, taking into consideration its complexity. As for any system, it has to be determined by the respective extensive and intensive parameters. This will imply to know the internal processes in the system. As it is known, only system with known structure could be described with the respective number of parameters, leading to the description of the internal interaction among the elements of the system. The complexity of the system can not be defined due to the lack of the knowledge of its structure.

Engineering science has been based on the canonic laws of physics and its derivatives in other branches of sciences. So, close link among these fields have been exercised in number of problems solving issues. To the less extend the engineering science has been integrated in the social aspect of the problems. The sustainability concept requires multidisciplinary approach in its full complexity. This is imposing understanding interactions among the different processes in the system. It is becoming apparent that single discipline has no possibility to coop with these kind of problems.

Engineering by its nature is the multidisciplinary field with enormous experience in treating problems with variety of options and complexity in the structure.

There are several ways in which the ideas of sustainable development are presented and interpreted. Ecocentric interpretations indicate a conspicuous degree of reference to the ecosystems. Anthropocentric interpretation tends to put humans at the center of the issues. Others such as biocentric interpretation focus on the protection of the elements of the biosphere. Our main concern in this general setting is energy and whatever we do for and with energy may be reflected in any of the above mentioned perspectives and interpretations. While this concern for sustainability in the energy sphere should include considerations at the global level, more importantly, there is the need to look at it as a regional issue in the overall global scenario. There are several perspectives on this. Figures 1 and 2 attempt to distinguish sustainable and unsustainable development.

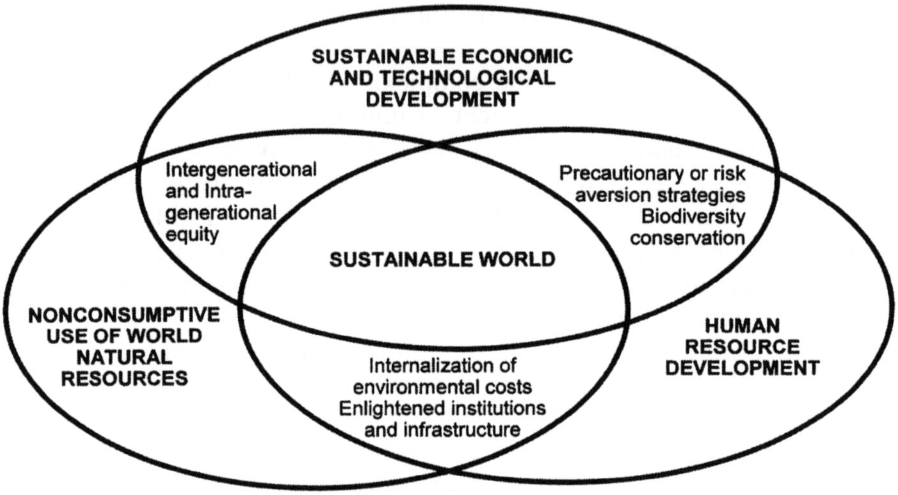

Figure 1. Sustainable development.

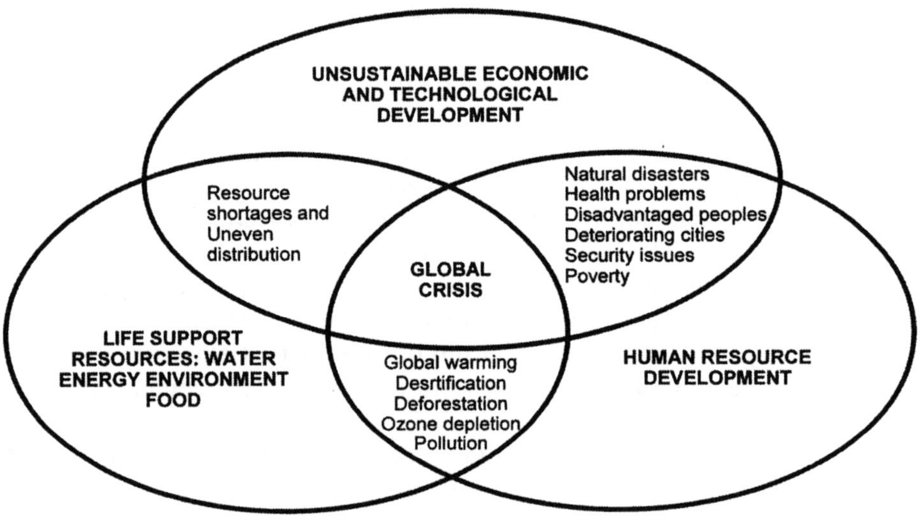

Figure 2. Unsustainable development.

3.2 Paradox of Sustainability

The notion of sustainability has generated tremendous interest and an avalanche of publications [12,13]. Some of them have not agreed in the definition of sustainability. Most definitions of sustainability are derived so that suit their particular applications, and often use term with no specific evidence and recognition of the exact meaning being implied.

The scientific analysis applied to the sustainability concept turnout to be either redundant, or ambiguous. There have been a number of attempt to focus on some of the problems and controversies of the goal of sustainability that emerge once it is treated within the frame work of system analysis.

The renewal cycle has been recognized in many dynamic systems and the cyclic pattern in life histories of complex systems. Cycles have been observed in numerous systems of very different nature. This cyclic nature of the development is contradictory to the goal of sustainability, which is generally aimed to preserve and maintain a certain state or function. In all renewal cycles it is assumed that a system goes through a series of stages starting from growth, followed by conservation then release and finally renewal. Within the framework of the renewal cycle, sustainability may be interpreted as the goal of breaking the cycle of extending a certain stage in the system life pattern. In this interpretation of the renewal cycle the sustainability concept will contradict with renewal. But if the phase of release of system cycle is the end, the collapse of the system is expected leading to the change of the system structure. This imply that the system function is modified. There are number of examples which prove this renewal cycling concept. Bankruptcy of a company when employees are laid off, and assets are sold (released) is the end of the company. It comes when this socio-economic system is no longer sustainable, it can no longer extend the conservation stage. The component (human and material resources) may be recombined in the form of another company (renewal), but that will be a different system.

Another aspect of the sustainability concept is the hierarchy of systems. If the renewal cycle is adapted, it will allow for readjustment and adaptation. But only next hierarchical level will benefit from this adaptation. Therefore, for a hierarchical system to extend its existence, to be sustainable, its subsystem needs to go through renewal cycle. In this way, death of subsystems contributes to sustainability of the super-system.

The argument presents here are reflecting different concepts of the approach to sustainability. This implied that the system approach to the sustainability concept is immanent to renewal cycle and hierarchical structure, what leads to the paradox of the sustainability concept.

4. SUSTAINABILITY ASSESSMENT

Sustainability development is a complex issue, which represents a multidimensional, multidisciplinary and complex structure of the system.

Measuring sustainability is a major issue as well as a driving force of the discussion on sustainability development. Developing tools that reliable measure sustainability is a prerequisite for identifying non-sustainable processes informing design-makers of the quality of products and monitoring impacts to the social environment. The multiplicity of indicators and measuring tools being developed in this fast growing field shows the importance of the conceptual and methodological work in this area. The development and selection of indicators require parameters related to the reliability, appropriateness, practicality and limitations of measurement.

In order to cope the complexity of sustainability related issues for different systems the indicators have to reflect the wholeness of the system as well as the interaction of its subsystems. Consequently, indicators have to measure intensity of the interactions among elements of the systems and system and its environment. In this view, there is a need for the indicator sets related to the interaction processes that allow an assessment of the complex relationship of every system and its environment. This will imply that complexity indicators will be defined reflecting links among internal parameters and external parameters of the system. This may be interpreted in the thermodynamic vocabulary as the intensive and extensive parameters of the system.

The example of energy system can be used to demonstrate the complexity of sustainability in the definition of sustainability indicators. As regard energy systems the attention can be focused on two approaches: one, based on complex system assignment to the energy system, and second, based on the conceptual options of the system. The complex system approach will see energy system as the entity changing in time as the result of internal and external interactions. This will lead to the use of indicators defining the processes of interaction. With this type of indicators the system will be described in the time scale, with the individual values of indicators and their trend. In this case the dynamic character of the interactions between the complex system and its elements may be followed. In particular, the social/nature interactions may be monitored and respective actions indorsed in due time. As regard the energy system, the indicators for complex system will make visible social and environmental changes before they become irreversible. This means that potential hazards could be detected and corrective action's taken in due time.

The conceptual approach to energy system indicators will describe potential options of the energy system. As the example, we can see the energy system with different energy sources with options reflecting potential situations imposed by the different aspects of the sustainable assessment. Also, indicators for the conceptual approach will use integral parameters common to the individual elements of the system. The conceptual approach is more suitable to the sustainability assessment of the energy system design. It may be used for the concept assessment It may be used for the energy system concept assessment. With respective indicators the energy system design could be compared, taking into consideration different aspect of the design. In compliance's with the concept of "natural capital" the indicator are defined to measure natural resources, environment capacity, social life quality and economic benefits of the specific options. It should be emphasized that indicators are not necessary numerical quantity. This will imply the need for the procedure, which will comply with the differences in the indicators measuring scale

The effective indicator has to meet characteristics reflecting a problem and criteria to be considered. Its purpose is to show how well system is working. In the case there is a problem an indicator has to indicate its origin and direction to take to solve the problem. Indicators are strongly dependent on the type of the system they monitor.

It is known that any numerical number, semantic expression or mathematical sign is information. Examples of these type of information can be found in any reading of temperature or other time dependent variables describing state of the system. Information is also parameters used in the design of respective system. Also, a positive or negative sign of the variable is also the information.

Collecting information and its processing will convert them in data. So, data represent agglomerated information, which are partially or finally processed. Examples of data can be find as a parameter, which describe evaluated information to be used for the specific purpose. In this respect the average inlet temperature of cooling water in the condenser is data obtained by the averaging procedure adapted for this purpose. Also, heat transfer coefficient used in the design of condenser is the data obtained by the experimental procedure for the heat transfer evaluation.

In order to use the data for the assessment of the respective system, it is necessary to convert them into the indicator. So, the indicator represents the measuring parameter for the comparison between the different states or structure of the system. As example, the efficiency of the system is an indicator for the quality of energy use in respective system. Also, we can evaluate different structure of the systems by the indicator representing respective entities of the system. In this direction is the assessment of intelligence use in the improvement of the system compatibility with its surrounding is measured by the respective indicators.

In order to quantify criterions for the sustainability assessment of any design of energy system element indicators are defined to meet this requirement. In this respect, the efficiency of resources use and the technology development is of the fundamental importance [14,15,16,17]. It is obvious that the efficiency of energy resource use is a short term approach which may give return benefit in the near future. As regard the technology development, the long term research and development is needed. In some cases it will require respective social adjustment in order to meet requirements of the new energy sources.

4.1 Definition of Sustainability Criteria

The criteria for the energy system sustainability assessment have to reflect a following aspects, namely: resource aspects, environment aspects, social aspects and economic aspects. In this respect, the sustainability assessment of the energy system will comprise evaluation of those parameters which are the reflection integral concept of the sustainability. As any other complex system the energy system is defined with the constrains which imply its function, technology, geography, property and capacity.

4.2 Measuring Sustainability

Measuring sustainability is a major issue as well as a riving force of the discussion on sustainability development. Developing tools that reliable measure sustainability is a prerequisite for identifying non-sustainable processes informing design-makers of the quality of products and monitoring impacts to the social environment. The multiplicity of indicators and measuring tools being developed in this fast growing field shows the importance of the conceptual and methodological work in this area. The development and selection of indicators require parameters related to the reliability, appropriateness, practicality and limitations of measurement.

In order to cope the complexity of sustainability related issues for different systems the indicators have to reflect the wholeness of the system as well as the interaction of its subsystems. Consequently, indicators have to measure intensity of the interactions among elements of the systems and system and its environment. In this view, there is a need for the indicator sets related to the interaction processes that allow an assessment of the complex relationship of every system and its environment. This will imply that complexity indicators will be defined reflecting links among internal parameters and external parameters of the system. This may be interpreted in the thermodynamic vocabulary as the intensive and extensive parameters of the system.

As regard energy systems the attention can be focused on two approaches: one, based on complex system assignment to the energy system, and the second, based on the conceptual options of the system. The complex system approach will see energy system as the entity changing in time as the result of internal and external interactions. This will lead to the use of indicators defining the processes of interaction. With this type of indicators the system will be described in the time scale, with the individual values of indicators and their trend. In this case the dynamic character of the interactions between the complex system and its elements may be followed. In particular, the social/nature interactions may be monitored and respective actions indorsed in due time. As regard the energy system, the indicators for complex system will make visible social and environmental changes before they become irreversible. This means that potential hazards could be detected and corrective actions taken in due time.

The conceptual approach to energy system indicators will describe potential options of the energy system. As example, we can see the energy system with different energy sources with options reflecting potential situations imposed by the different aspect of the sustainable assessment. Also, indicators for the conceptual approach will use integral parameters common to the individual elements of the system. The conceptual approach is more suitable to the sustainability assessment of the energy system design. It may be used for the concept assessment It may be used for the energy system concept assessment. With respective indicators the energy system design could be compared, taking into consideration different aspects of the design. In compliance with the concept of "natural capital" the indicators are defined to measure natural resources, environment capacity, social life quality and economic benefits of the specific options. It should be emphasized that indicators are not necessary numerical quantity. This will imply the need for the procedure, which will comply with the differences in the indicators measuring scale.

The effective indicator has to meet characteristics reflecting a problem and criteria to be considered. Its purpose is to show how well system is working. In the case there is a problem an indicator has to indicate its origin and direction to take to solve the problem. Indicators are strongly dependent on the type of the system they monitor.

It is known that any numerical number, semantic expression or mathematical sign is information. Examples of these type of information can be fined in any reading of temperature or other time dependent variables describing state of the system. Information is also parameters used in the design of respective system. Also, positive or negative sign of the variable is also the information.

Collecting information and its processing will convert them in data. So, data represent agglomerated information, which are partially or finally processed. Examples of data can be find as a parameter, which describe evaluated information to be used for the specific purpose. In this respect the average inlet temperature of cooling water in the condenser is data obtained by the averaging procedure adapted for this purpose. Also, heat transfer coefficient used in the design of condenser is the data obtained by the experimental procedure for the heat transfer evaluation.

In order to use the data for the assessment of the respective system, it is necessary to convert them into the indicator. So, the indicator represents the measuring parameter for the comparison between the different states or structure of the system. As example, the efficiency of the system is the indicator for the quality of energy use in respective system. Also, we can evaluate different structure of the systems by the indicator representing respective entities of the system. In this direction is the assessment of intelligence use in the improvement of the system compatibility with its surrounding measured by the respective indicators.

In order to quantify criterions for the sustainability assessment of any design of energy system element indicators are defined to meet this requirement. In this respect, the efficiency of resources use and the technology development is of the fundamental importance. It is obvious that the efficiency of the energy resource use is a short term approach which may give return benefit in the near future. As regard the technology development, the long term research and development is needed. In some cases it will require respective social adjustments in order to meet requirements of the new energy sources.

Potential improvement of the energy conversion process is a driving force for its development. In the assessment of the conversion process a promising tool is the exergy analysis of the energy system. The exergy analysis is based on the maximum potential availability and its use for the assessment of the conversion process. By definition the exergy is a parameter for the validation of the efficiency of energy conversion process and system. Taking into account the law of thermodynamics the technology improvement appears as a significant factor responsible for an entropy change in energy system. The application of the principle of Carnot therefore allows to determine an absolute limit to any transformation of the deposit of free energy.

4.3 Sustainability Assessment Indicators

In order to facilitate the tool for sustainability assessment the indicators are used as parameters for the measurement of sustainability. In this respect, attention is focused on the resource indicators and environment capacity indicators. Each of this entities has to be determined with specific parameters which can be used to determine the characteristic indicators for the assessment of the respective entities. The characteristic indicators are:

a.) Possible Change Indicator

Possible change indicator is characterising changes, which are determined by the maximum change technically possible. It reflects the difference between state of the entity with maximum availability. If applied to the energy resource, it is a difference between known resources exploitable with known technology and current resource to be obtained with present technological capacity. As example, it can be shown that coal resource in 1980 have been 2000 Mtoes and in 1985 2300 Mtoes. Maximum potential resources to be used without effecting the available resources is 300 Mtoe. The indicator for the possible change is defined as the maximum potential resources and the average value for the period under consideration. So that the Possible Change Indicator (PCI) is 0.13. Possible change indicator is a function of the technology development in the resource discovery. This implies that there are two means for the increase of the resources, namely: the additional investment for the new discovery and the new technology for the resource exploration.

b.) Current Change Indicator

Current change indicator is defined as the resource consumption change of the respective resource. It can be defined as the difference in current consumption change in the reference time period. Applying the measurement of this indicator to the period 1980 – 1982, it can be seen that Current Change Indicator (CCI) for coal is 0.467. It is known that the current consumption of the energy resources strongly depend on the efficiency of energy use. Possible change in the efficiency of energy use includes the efficiency of the primary energy source conversion and second, the efficiency of the finale energy use.

c.) Resource Indicator

By definition, the Resource Index (RI) is the ration between the Current Change Indicator and the Possible Change Indicator. In essence it will represent the qualitative measure of the resource depletion. Also, it may be interpreted as the resource scarcity change due to the resource depletion. For the case of coal resource the Resource Indicator is 0.065.

d.) Financial Effect of Resource Use

The resource indicator is reflecting material balance of the respective primary energy resource. In order to use it for the quality assessment of the primary energy use, it has to be connected with the financial effect resulting form its use.

5. REFERENCES

1. Sustainable Energy Strategy, National Energy Policy Plan, Department of Energy Organization Act., Recommendation to the United Nations Commission on Sustainable Development, ISES NGO Renewable Energy Institute, 1995.
2. Report of The United Nation Conference on Environment and Development, Vol. 1, Chapter 7, June, 1992.

3. Agenda 21, Chapter 35, Science for Sustainable Development, United Nation Conference on Environment and Development, 1992.
4. Declaration of the Council of Academies of Engineering and Technological Sciences.
5. The Earth Chapter: A Contribution Toward its Realization, Franciscan Center of Environment Studies, Roma, 1995.
6. Jenkinson, C.S., The Quality of Thomas Jefferson's Soul, White House, Library, 1978.
7. Medows, D. H., Meadows, D.L, Randers, J., Behrens III, W.W., The Limits of Growth, Universe Book, New York, 1972.
8. Report of The United Nation Conference on Environment and Development, Vol. 1, Chapter 7, June, 1992.
9. Becker, E., Jahn, T., Stiess, I., Wehling, P., Sustainability: A Cross-Disciplinary Concept for Social Transformations, MOSCT Policy papers 6, UNESCO September, 1997.
10. Pearce, A., Turner, R.K., Economics of Natural resources and the Environment, Harvester Wheatsheeaf, London, 1990.
11. Afgan, N.H., Al Gogaisi, D., Carvalho, M.G., Cumo, M., Sustainable Energy Development, Renewable and Sustainable Energy reviews, 2, pp. 235-286, 1998.
12. Voinov, A.A., Paradox of Sustainability, Institute for Ecological Economic, Solomons, USA, 1995.
13. Constanza, R., Voinov, A., Interaction Between Economic and the Environment form the Point of View of Sustainable Development, Edited: Koptyug V.A.,Uppenbrink M., Sustainable Development of Lake Baikal region, NATO ASI Series Partnership Sub-Series 2, Environment,Vol. 6, Springer Heidelberg,pp.33-58, 1995.
14. Afgan, N.H., Carvalho, M.G., Sustainability and Heat Exchanger Design, Heat Transfer Enhancement of Heat Exchangers, Ed.: S. Kalkac et al., Kluwer Academic Publisher, ISBN-07923-5637-3, 1999.
15. Afgan, N.H., Editorial: Sustainability in Thermal Engineering, Heat Transfer Engineering, Nov. 1999.
16. Hovanov, N.H., Analysis and Syntheses of Parameter under Information Deficiency, (in Russian), St. Petersburg, Publisher: St. Petersburg University, ISBN 5-288-01533-3, 1996.
17. Hovanov, N.H., Kornikov, V., Seregin, I., Quantitative Information Processing in DSS ASPID-3W for Complex Objects Estimation under Uncertainty, Proceeding of the International Conference "Information and Control", St. Petersburg, pp. 808-816, 1997.

CHAPTER 3

SUSTAINABLE ENERGY DEVELOPMENT

1. INTRODUCTION

In order to reach the goals indicated by the sustainable energy development the efficiency in the energy conversion use has to meet several criterions [1]. The potential for the efficiency improvement is generally underestimated. Most of the energy conversion systems consider the efficiency improvement as a separate process and their analysis reflects only the potential improvement of the process but not the potential for the efficiency improvement obtained by an exergy analysis of the energy system. Fossil fuel energy resources use is mostly conversion to heat by the combustion processes. Since the combustion process is taking place at temperatures between 900 – 1300°C and over 40% of heat is used a low temperature heat, it is indispensable to take into consideration the thermodynamic assessment of the efficiency in order to bring in line energy conversion processes and energy demand to obtain the optimum fuel utilization.

In the definition of sustainability it is of substantial importance to envisage its broad aspect which composes versatility of the components to be taken in consideration [2,3,4]. In this respect, wholeness of sustainability has to include definitions of those components which are linked to specific parameters to be taken into a consideration of the assessment of sustainability of specific situations in global, regional and local environments. There is a number of characteristic entities which will be used to define the wholeness of the sustainability: life diversity, natural resources, environment capacity, population increase, social disturbances and ethic changes.

It is out of the scope of this writing to dwell on all characteristic entities for sustainability definition. Even, the energy is one of the essential commodities required for human life and is affecting the sustainable wholeness in its total change [5]. For this consideration we will focus our attention only on those characteristic entities which are in direct correlation with energy sustainability measurements. In this group are included: i.) natural resources; ii.) environment capacity; iii.) population increase.

Each of these entities has to be defined with specific parameters which can be used to determine characteristic indicators for global assessment. The possible use of indicators is characterizing the changes which are determined by the maximum values technically feasible. It is reflecting the difference between the state of the entity with the maximum availability and respective current state of the entity. If applied to the individual energy resources there is a difference between known maximum exploitable resources with known technologies and current resources to be obtained with present technological capacity. Possible changes are functions of the technological development in the resource discovery. This implies that there are two means for the increase of the resources, namely, by additional capital investment for the discovery and by the new technology for the resource discovery.

In the assessment of sustainability, the current consumption change has to be taken into consideration which is reflecting the current consumption change in the reference time period. In order to form some kind of resource indicator for sustainability measurement a ratio between the current change and the maximum potential change has to be established. Its trend will give the measuring parameter for the resource depletion in time. It is known that the current consumption of the energy resources strongly depends on the efficiency of their use, which may be classified in two groups. The first one is the possible efficiency increase due to the change in the efficiency of primary energy source conversion and the second one is due to the change in the efficiency of the final energy use.

2. NON-RENEWABLE ENERGY SOURCES

A number of authoritative studies have presented forecasts for the energy supply in the 21st century. Conclusions drawn from this analysis have become a driving force for the development of the plan for the sustainable energy supply system. Even if there is a number of options taken into consideration, common issues are the following:

2.1 Prevention of the Energy Resources Depletion with Scarcity Index Control

Whichever the scarcity model is used, the energy resource scarcity is in direct relation to the social production output. In this respect, the efficiency of resources use and technology development are of fundamental importance. It is obvious that the efficiency of the energy resource use is a short-term approach, which may give a return benefit in the near future [6]. As regards the technology development, a long term research and development is needed. In some cases it will require respective social adjustments, in order to meet requirements of the new energy sources.

The availability of energy resources is limited by two factors: capital to be invested in prospecting of new resources and prospecting technologies for energy resources.

From recent experiences it was learned that there is a direct correlation between the capital invested in prospecting and the amount of the available reserves. It was proved that a fixed amount of 18 \$/t is needed for new energy reserves. In many developing countries this is a limiting factor for the availability of energy resources.

The prospecting technology is composed of three phases. The geological subway based on the real prospecting and respective diagnostic techniques for electromagnetic waves detection. The resolution of the instrument employed is one of the limitations, and it is under consideration for further development.

The second phase of prospecting technology is related to software for the design of the resource body based on the ultrasonic scattering or earthquakes generated by local explosions. The main limitation in the development of new software is the speed and memory size of computers. It can be expected that with the further development of computer technology this problem will be overcome. Also, new numeric schemes will substantially contribute to the accuracy and time expectation for the prediction of the size of resources body.

From the beginning of the energy resources exploration, the drilling technology was the limitation in the achievement of new resources. The development of drilling technology has marked a direction for the discovery of new resources. A recent example with a new drilling technology has led to a gas resource in the Mexican bay. Also, North Sea gas resources are the result of a new off shore drilling technology. The same has been proved in the discovery of new gas resources in Algeria. The deep sea drilling has become one of the global issues, which may remove the scarcity problems of energy resources for the next few centuries. It should be mentioned that 2/3 of the earth surface is covered by deep sea so that the breakthrough in deep-sea technology may lead to a substantial change in the energy resource picture in the world.

2.2 Efficiency Assessment

The potential improvement of the energy conversion process is a driving force for its development [7]. In the assessment of the conversion process a promising tool is the exergy analysis of the energy system. The exergy analysis is based on the maximum potential availability and its use for the assessment of the conversion process. By definition, the exergy is a parameter for the validation of the efficiency of the energy conversion process and system. Taking into account the law of thermodynamics, the technology improvement appears as a significant factor responsible for an entropy change in the energy system. The application of the principle of Carnot therefore allows to determine an absolute limit to any transformation of the deposit of free energy [8,9].

Following the first energy crisis, many countries have organized an energy efficiency assessment campaign with the aim to improve the efficiency and gain saving which has contracted the increase of energy price. It was proved that this approach has resulted in the increase of efficiency of energy use between 10 – 20% in a number of European countries. The main emphasis has been given to the evaluation efficiency of different technologies and utilizations of energy.

It is of great importance the effort directed to the evaluation of the technological processes for energy saving [10,11]. Also a new development of products is recently under consideration for the minimum use of energy. In accordance with one of the criterions for the sustainable development, products have to meet the requirement related to the minimum use of energy. A favourable example for this achievement is the development of a new lighting system with fluorescent lamps and which, in comparison with traditional bulbs, have a saving of about 40%.

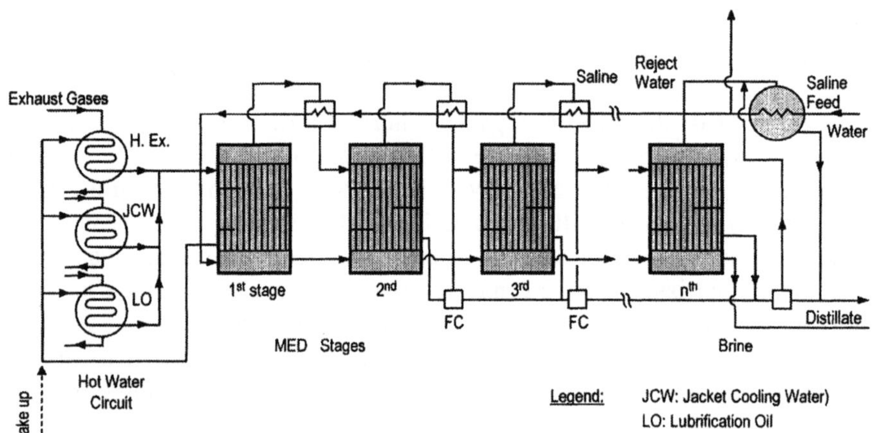

Figure 1. Schematic representation of cogeneration in desalination plants.

Cogeneration of heat and electricity is one of the potential means to improve the efficiency of the energy resource utilization. Cogeneration plants in conjunction with the desalination in regions where water and energy are needed are an example. Figure 1 shows a schematic representation of cogeneration in desalination plants [12].

Recent projects with gas fired cogeneration plants have demonstrated an extremely high efficiency [13]. The increased gas resources may lead to further development of highly efficient power plants for electricity production. The cogeneration will play a special role in the development of new energy systems.

2.3 Clean Air Technology Development

The combustion process is an irreversible thermodynamic process with a high degree of availability losses in the energy conversion cycle. In this respect there is a potential opportunity to increase the energy conversion efficiency by improving the combustion process. There are a number of potential combustion technologies which might lead to an efficiency increase of the combustion process. Among those there are:

a. Catalytic Combustion

The low temperature catalytic combustion of lean natural gas mixtures represents an effective method for heat generation [14]. Coexistence with reactant catalysts enhances the chemical reaction but is stechiometrically independent of the reactant. Among the processes of catalysis there is the absorption into the catalysis reaction at the catalyst surface and the liberation of the chemical products. Zeolite is a catalyst widely used in chemical industry.

Detailed behaviour of the catalyst has not been fully understood. In particular it is expected that the catalytic combustion may lead to an efficient use of the fuel cell. Figure 2 shows a schematic representation of the catalysis of platinum in a fuel cell.

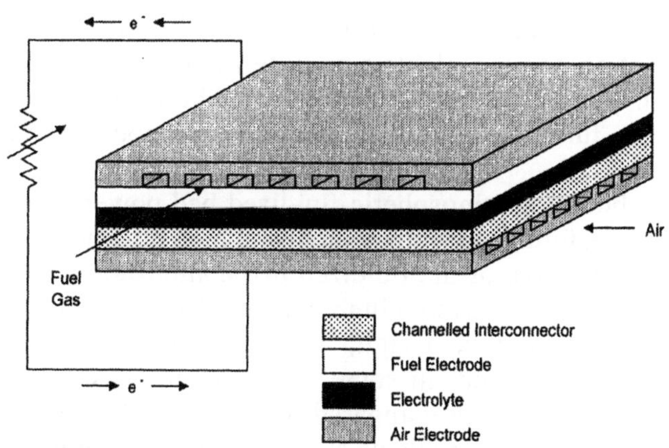

Figure 2. Schematic representation of the catalysis of platinum in a fuel cell.

The catalysis mechanism at the interface between electrode and electrolyte ensures the electron transfer from the input hydrogen molecule to the electron metal. The search for low cost alternatives has not been very successful but lately the good performance of some of active composition of La, Ni, Co, O (LSNC powder) leads to promising expectation.

b. Fluidized Bed Combustion

A recent progress in the fluidized bed combustion has led to substantial developments of this new energy system [15]. In the combustion in fluidized beds the coal is depressed in a mass of its ashes and absorbent lime and the process is developed at a temperature of 850°C. There are two types of boilers: bubbling and circulating (Figure 3).

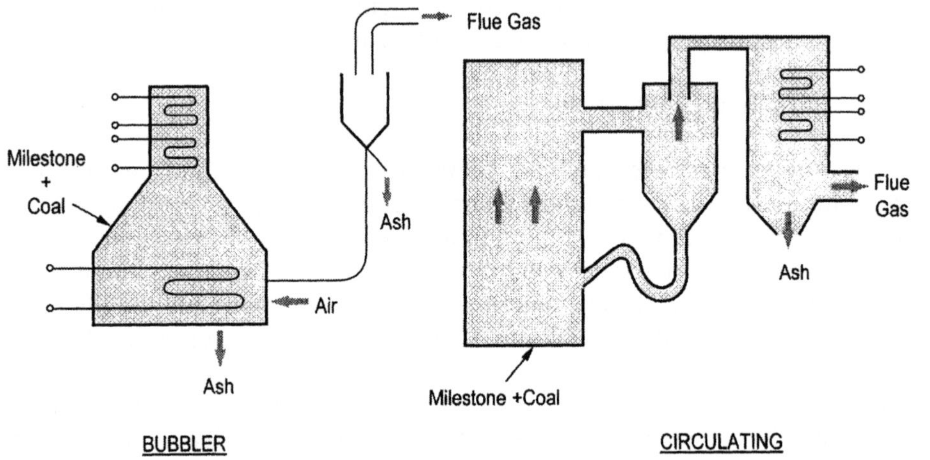

Figure 3. Types of boilers: bubbling and circulating.

The bubbling alternative offers a good thermal design. In principle, this was a clean option for the electricity generation with medium and good quality of coal. The energy efficiency of a bubbling boiler in the Rankine cycle with steam turbines, is similar to that of the conventional pulverized coal power plant.

A 350 Mwe bubbling atmospheric fluidized bed power plant is an option with a very good performance with medium and high quality coal.

The second alternative of the fluidized bed combustion power plant is the circulating fluid design, offering a high degree of operating flexibility in coal quality use. It is a complex design, which includes fuel chambers, large cyclone, recovery boiler and, in some cases, outer ash cooler. This option reduces the energy efficiency compared to present pulverized coal plant. A factor contributing to these problems significantly is the high electric energy consumption in auxiliary services, particularly for the ventilator for recirculation. The efficiency of the existing circulating fluidized bed plants is about 30%. A 250 Mwe plant is in operation with low quality coal. The use of circulating fluidized bed boiler technology is rapidly increasing due to the ability to burn low grade coal, while meeting the required NO_x, SO_x and particulate emission requirements.

The pressurized fluidized bed combustion boiler is an option offering a 10% increase in efficiency over conventional pulverized coal fired plants. It is a compact plant with moderate specific investment using high quality material and it is conceived for medium and good quality coal.

c. Low NO_x Burners

The present, advanced energy technology is focused to reach further improvements in the emission control [16]. In principle there are two approaches: the first one, by reorganization of combustion processes in the burners and the second one, by post-combustion processes in the furnace.

In order to minimize SO_x, NO_x and particulates emissions a new burner design is envisaged to meet the requirements for minimization of initial NO_x formation. It is expected that the new burners will reach 370 – 490 mg/Nm^3 in properly designed new furnaces [17], which is the limit for present emission control.

Further NO_x reduction can be achieved through furnace staging. Here the boiler combustion zone is opened close to the stechiometric chemistry condition and the balance of air is added in the upper furnace through an overfire airport. NO_x emission can be lowered through post-combustion technologies such as selective catalytic reduction. NO_x is reduced with molecular nitrogen and water and by reduction with ammonia in presence of a catalyst.

Numerical modeling of the processes in combustion chambers has become important in design and analysis of tools [18,19] for improving air distribution in power plant burners. Numerical modeling allows the analysis of designs for optimal modifications of turning vanes, flow splitters, perforated plates and burner shrouding.

Also, numerical models of boiler furnaces [20] are available as computational fluid dynamic software which allows practical analysis of power plant furnace behaviour with minimum emissions of SO_x, NO_x and particulate. Retrofitting of existing power plants with advanced combustion technologies will lead to substantial increases of efficiency and minimize emission, to the environment [21,22]. Figure 4 shows typical low NO_x burner.

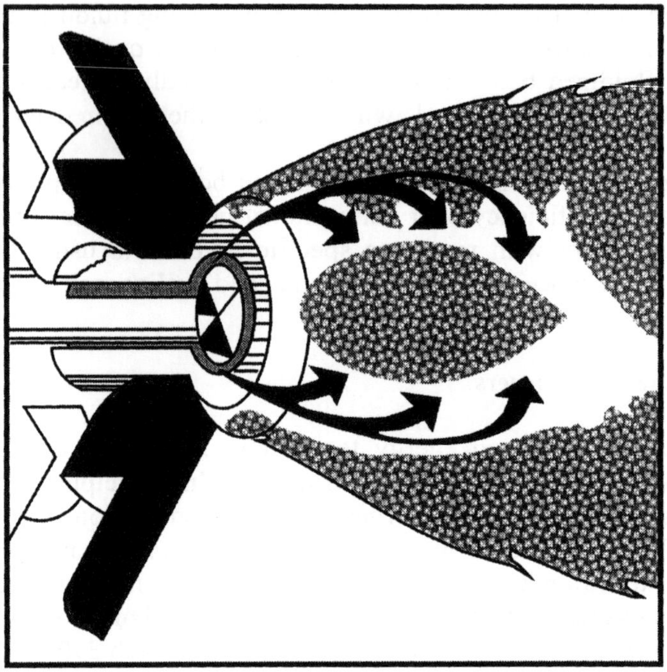

Figure 4. Schematic representation of new boiler design.

d. New Boiler Designs

Coal fired boiler power plants will continue to be one of the major contributors in the future. Modern pulverized coal-fired systems presently installed generate power at net thermal efficiency ranging from 34 – 37%, while removing up to 97% of uncontrolled air pollution emissions. A new generation of pulverized coal-fired boiler technologies are currently under development, which will permit a generating efficiency in excess of 42%. In this respect, further development is needed for the improvements to reduce the emissions and expande the operability. To achieve a high thermal, special attention has to be devoted to the load cycling operation. In this respect a low-emission boiler system in development is based on the diagnostic monitoring of process parameters and expert systems.

The development of high performance power systems is an ultimate goal for upgrading pollution control with corresponding combustion system improvements and the control of SO_x, NO_x through the implementation of new burner design and post combustion emission control. Also, the implementation of the boiler numerical codes for the determination of process parameters will be used as a tool for the efficiency control and early diagnostic function monitoring. This improvement in heat and mass transfer research [23,24,25,26] has substantially contributed to new boiler designs and will lead to the increase of availability of modern power plant systems [27]. Figure 5 shows the schematic representation of new boiler design.

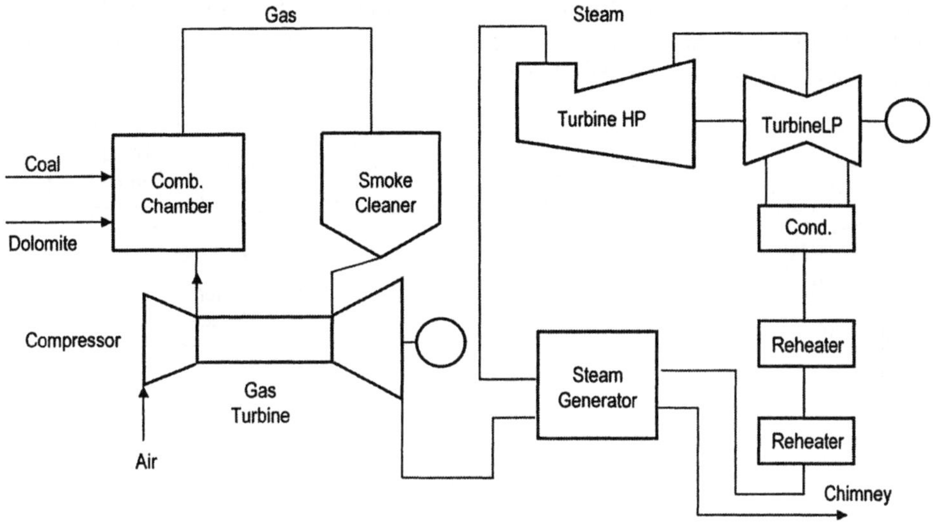

Figure 5. Expert system for fouling assessment.

2.4 Development of Intelligent Energy Systems

The recent development of artificial intelligence has opened the possibility to utilize those achievements in the energy sustainable development. There are three major paths, namely: expert system development in energy engineering; new control based on fuzzy logic and respective reasoning; intelligent thermal system design.

a. Expert System in Energy Engineering

The expert system development in energy engineering is focused in two directions: expert system for energy system design and knowledge-based for on-line diagnostic [28,29,30,31]. It has been shown that the expert systems for energy system design can be an efficient tool in selection, optimisation and assessment of power plant design. Also, expert system logic can be used in energy system planning, including optimization of the energy system, reflecting the potential use of renewable energy sources. An example of expert system use in the design of thermal equipments is demonstrated by the heat exchanger design [32]. Further developments of knowledge - based system for design of energy system will promote the increase of efficiency and reliability.

The knowledge-based system for the fault diagnostic in energy systems has proved to be a powerful tool for the evaluation of system parameters in order to forecast a potential malfunction of system elements. Figure 7 shows a schematic representation of expert system for fouling assessment.

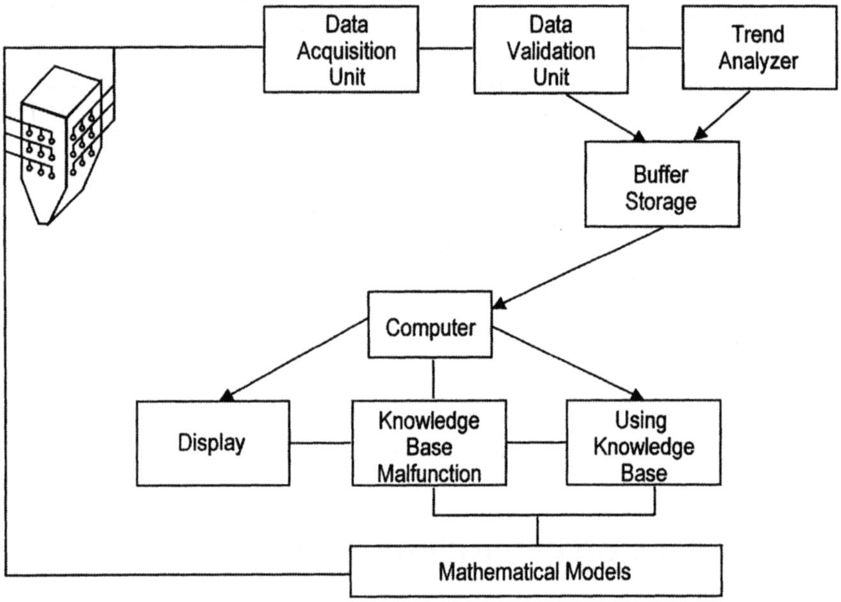

Figure 6. Block diagram of fuzzy control systems.

There are several attempts which have proved the possibility of knowledge-based systems in the diagnostic of thermal power plants. The efficiency monitoring and respective logistic evaluation of diagnostic parameters has been demonstrated to be a good and reliable tool for the advanced diagnostic of operational deficiency. The boiler fouling and tube leakage knowledge - based system prototypes demonstrated the possibility of the detection of processes leading to the degradation of power plant efficiency [33]. The diagnostic systems are based on the on-line monitoring of diagnostic variables and their fuzzyfication for the reasoning retrieval of the cases representing diagnostic results.

b. Fuzzy Logic Control

The new fuzzy logic control system is demonstrated to be a qualitatively efficient system for the on-line control of energy systems [32]. While similar model-based control systems designs are trial and error, the knowledge-based controller is "ad hoc" at present time. A gap exists between solid theoretical results such as stability, controllability, etc. A real-time implementation of intelligent control systems uses fuzzy logic, neural networks, generic algorithms, expert systems, etc. A common definition of fuzzy control system is that it emulates a human expert. Under this situation, the knowledge of human operator would put in form of a set of fuzzy linguistic rules. These rules would produce an approximate decision, like a human would. Figure 7 shows a block diagram of fuzzy control systems.

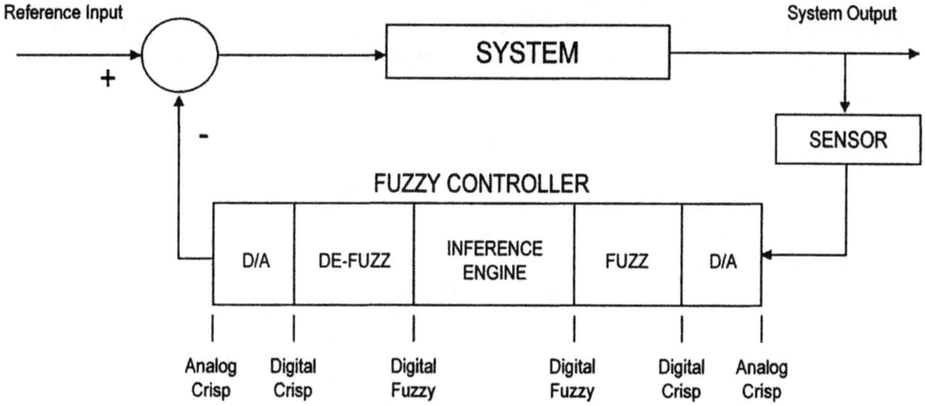

Figure 7. Intelligent energy system.

c. Intelligent Energy Systems

The present world has learned that some of the basic assumptions are not valued and so requires a new approach for adjustment to its future development. There have been world scale meeting emphasising the need for the economic order to meet contemporary development within the limits recognised by irreversible changes in energy resources and environment capacity.

New measurements called "Indicators of sustainability" are designed to provide information for understanding and enhancing the relationships between the economic energy use, and environmental, and social elements inherent in long-term sustainability.

There are different terms used in the consideration of the product design. The "clean" design is used in the widest sense of all management as well as technical decisions related to specification, planning and development of products not just technical processes designed by engineering. The "eco"-design refers to the processes of systematic incorporation of environmental life cycle consideration into product design.

"Intelligent product" design comprises the specifications reflecting resource life cycle, environmental cycle, product life cycle, end of life and clean design. The generic design procedure to be adopted for the intelligent product design of the thermal equipments require the definition of indicators for the assessment and optimization of the specific design.

In order to provide the design criterions reflecting complex requirements imposed by the intelligent design, it is necessary to define the respective indicators to be used in the evaluation of the specific design of thermal equipment [33,34]. These indicators should be based on the optimization of the efficiency of respective thermal equipment, resource use assessment and validation, environment capacity use and degradation, modular structure with multipurpose elements, end of life assessment and economic justification of specific designs. Figure 8 shows schematic presentation of Intelligent energy system.

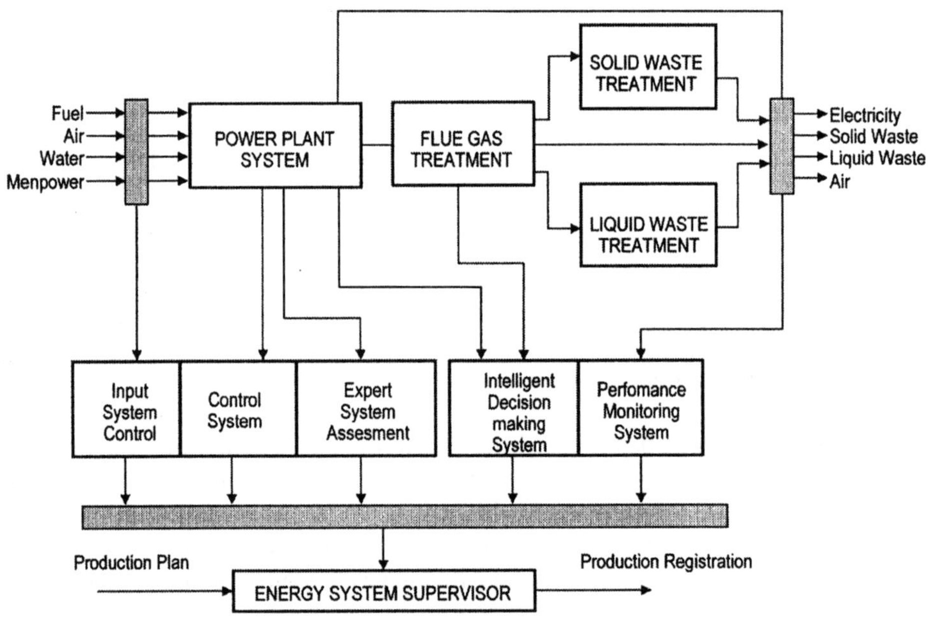

Figure 8. Schematic representation of Solar power plant.

In order to evaluate the validation of the indicators the thermal equipment will be used. In this respect, the criterions will be adapted for the specific thermal equipment. It is aimed to develop the algorithm with the indicators to be used for the assessment of design. The assessment will be made for a number of selected products presently on the market to be evaluated within the frame of the criterions for intelligent design.

3. NEW AND RENEWABLE ENERGY SOURCES (NRES)

Besides the possibility offered by the efficiency improvement of processes in power generating units, there is a great challenge to increase the efficiency of energy systems by introducing energy mixed systems [35, 36]. The energy system includes power or heat generating units, energy transport systems, energy storage systems and energy end-use systems. From the energy resources point of view there is a number of options which might be taken into consideration for the optimization of the energy system. In this respect, high interest is shown to the potential use of the renewable energy resources as power or heat-generating unit in the energy system. The immanent advantages of the renewable energy resources due to their availability and low cost impact, are promoting the renewable energy source to be included in the energy system.

Renewable energy sources by definition, meet the requirements of sustainability. It is therefore expected that the long-term energy strategy will rely on renewable energy resources. The total availability of the renewable energy resources is very large. This picture reflects the presently available technologies in the field of renewable energy resource use and exploitation [37,38]. Very promising alternatives are envisaged with the promotion of new technologies under development.

3.1 Solar Energy Resources

Solar energy can be exploited in three main modes:

- by enhanced absorption of solar energy in collectors, which provide low-grade heat;
- by using reflecting devices to concentrate the solar energy in a heat carrier, which is then used to generate electricity;
- by converting sunlight directly into electricity.

Solar energy resources do not have clear limits. The annual influx on the Earth surface is 10 000 times as large as the current human energy consumption; the fraction reaching the land surface is 3 000 times as large and even 35% of this would make 1 000 times more energy than we demand today. As we can notice the resource of solar energy is huge but diluted. In the literature it is assessed that feasible tool use of solar energy from the technical standpoint is about [35]:

Thermal solar for	170 Mtoe/year
Decentralized electric solar for	450 TWh/year
Network electric solar for	230 TWh/year

In the local resources evaluation for these three solar systems one could take into a consideration its minimum and maximum capacity to be installed. From the present status of the development the following capacity can be taken into consideration

	Minimum:	Maximum:
Thermal solar	150 kW	80 MW
Decentralized electric solar	30 kW	5700 kW

If the mean insulation for the specific location is taken into consideration than the respective values for the land to be used will be obtained. For this purposes, it will be used q_R = 5.4 kWh / m^2 /day so that the following extension of land is required for the specific use of the solar energy:

	Minimum land[m^2/kW_e]
Thermal solar	20 - 35
Decentralized electric solar	36 -80

It should be mentioned that for insulation lower than q_R = 4 kWh / m^2 /day, it might be difficult to adopt the same method of validation.

Solar energy use is presently demonstrated in three options: solar thermal, solar photovoltaic and solar power plant. Solar thermal energy production plant has reached an industrial level and is available on the market in many countries[36,37,38]. There is a number of designs differentiated with respective efficiency and sophistication of the material used. Solar thermal option is mostly available for the hot water production units. It is also demonstrated for air heater and climatization units. Since it has reached maturity, it is not expected major breakthrough in this field, which might affect its potential use. Water heater units are available in the market with the following ranges:

	Temperature range [°C]
Unglazed flat plate collectors	< 40

Glazed flat plate collectors	40 – 100
Vacuum tube collectors	150 – 200

The solar thermal is anticipated to cost 8 – 9 UScents/GJ in the medium term. In the long term this could come to 4 – 6 UScent/GJ.

The solar photovoltaic system is also in its demonstration stage, with a number of various applications. It has been demonstrated in three levels, namely:

	Power
Ubiquitous solar cell	not fixed
Solar unit for electronic application	50W - 1 kW
Solar units for irrigation	5 - 60 kW

The first level is not an energy intensive application and has no significance for its consideration for the energy source strategy point of view. The second option is being used as the only energy source in the remote areas and has been demonstrated as a reliable energy source. It ranges in power production from 50 W to 1 kW and is commercially available at competitive prices. Figure 9 shows schematic representation of Solar power plant.

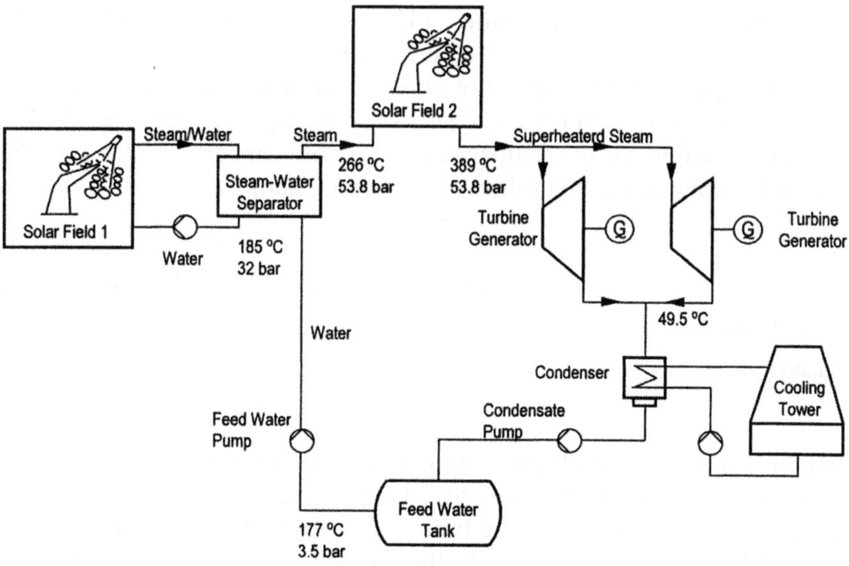

Figure 9. Minimum production temperature in a geothermal field that is required for different utilizations.

For the third option there has been a number of demonstration projects which have proved its feasibility for rural areas. Even, the energy available from the grid for many countries is lower than the energy produced by a photovoltaic unit, and the difference in energy may be compensated by the additional capital investment needed for a new grid construction for remote areas.

Solar power plants are in the stage of development. They are available as photovoltaic cell modular units and can be installed when the power demand requires a system augmentation. As demonstration, the modular power ranges between 50 kW and 1MW. Solar power plants with concentrators are still in the development stage and will not be considered in the technical assessment of the solar energy utilization [39,40].

Solar plant power 50 kW - 1 MW

Present estimate of photovoltaic solar plant electricity ranges from a cost of 23 to 33 UScent/kWh. It is expected that in the medium term this cost will be as low as 2.2 – 2.4 UScents/kWh [41].

3.2 Geothermal Energy Resources

Resources exploitable at current energy prices correspond to aquifers in narrowly localized volcanic zones [42,43]. Presently installed and in construction plants reach a total electrical capacity of 7 100 MW (high enthalpy energy). Low enthalphy hot water to be used directly for heating is estimated to about 13 Mtoe/year. These two groups of geothermal energy systems are specified by the temperature and flow capacity of the individual well. From the experience, the following limits are adopted:

	Min.temperature [°C]	Min.capacity [m^3/h]
High enthalphy geothermal	90	2900
Low enthalphy geothermal	35	1000

Particular resources evaluation will require a specific assessment of the respective aquifer volume in order to estimate uncertainty in the lifetime of the geothermal energy system to be built. It is expected that the volume of the aquifer should exceed several times the capacity needed for the respective system lifetime. In the local resource evaluation for the geothermal system it should be included also the land required for the brine to be deposited during the lifetime of the plant.

Geothermal utilization is commonly divided into two categories: electric production and direct application. Figure 10 shows the minimum production temperature in a geothermal field that is required for different utilizations.

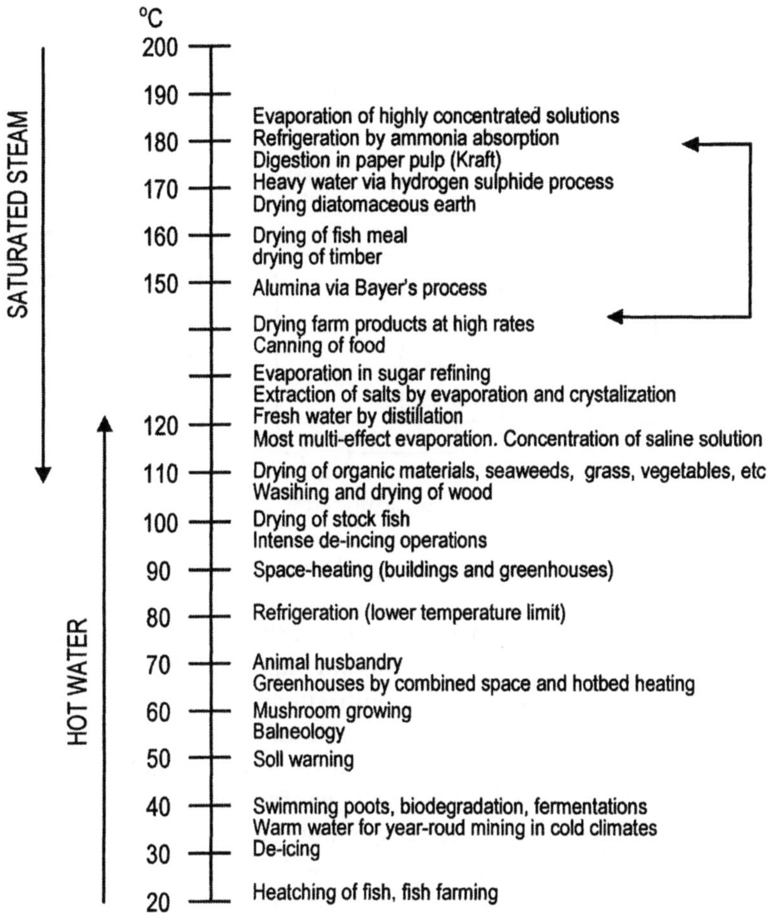

Figure 10. Schematic representation of direct steam geothermal plant.

Conventional electric power production is limited to fluid temperatures above 140°C, but a considerable lower temperature may be used with the application of binary fluids. Geothermal electric energy plants have reached their maturity and have proved to be very reliable sources of energy. They are used in 21 countries, with a total world installed capacity of 6017 MW distributed over 330 individual turbine generator units. The geothermal power plants are built in four versions, namely: direct steam plants, flash steam plants, binary plants and hybrid plants.

Direct steam plants are used with vapour dominated resources. Steam from production wells is gathered and transmitted via pipelines directly to a steam turbine. In most direct steam plants the capacity of the turbine is greater than 5 MW. Typical schematic representation of direct steam plant is shown on Figure 11.

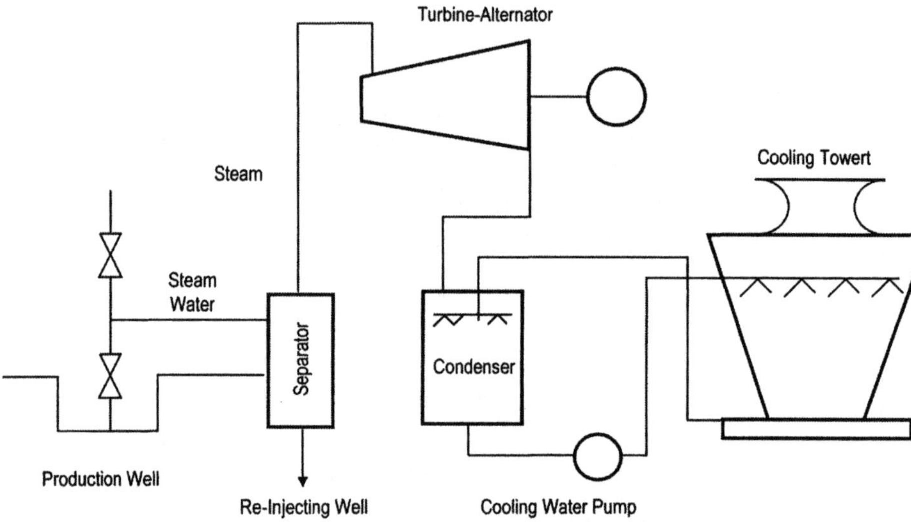

Figure 11. Schematic representation of typical biomass power plant

Flash steam production is used when the fluid, in the geothermal reservoir is pressurized hot water or a mixture of liquid and vapour at the well head. It is designed in two modifications, namely, with single flash steam unit and dual flash units. Single flash plant is a simple flash unit. The flashing process takes place between the reservoir condition and the power plant. The flash usually occurs in the well at the point where the geofluid pressure falls to the saturation pressure corresponding to the temperature and composition of the geofluid. A significant improvement in resource utilization can be achieved by adding a secondary flash process. Instead of being discharged, the liquid that is removed from the separator is subjected to another pressure drop in which releases additional steam. The lower pressure steam is admitted to the turbine at an appropriate stage and generates additional power.

Binary plants are used when it is not advisable to allow the geofluid to come in contact with the power production equipment, i.e., the turbine, for danger of scaling, corrosion or effects of large quantity of non-condensable gases. In basic turbine system, the hot liquid enters a vapour generator where it transfers heat to a secondary fluid. The working fluid is chosen to provide good thermodynamic matches to geofluid and obtain a high utilization efficiency, safety and economy.

Besides the basic binary cycle there are two cycles, which exhibit a higher efficiency: a dual pressure system and a dual-fluid system. In dual pressure binary plants the turbine receives the working fluid at two pressure levels. Either two separate turbines or a dual admission turbine may be used. Optimized dual-pressure binary plant can have 15-25% higher brine utilization compared to the basic binary plant geofluid temperature of 95-150°C. The dual fluid binary plant is a system with two binary loops, each of them with a particular fluid. The key for producing a high efficiency cycle is to couple the two loops by means of a recuperator in which the heat otherwise wasted, from the upper loop is transferred to the lower loop in sufficient quantity to provide the evaporation for the lower cycle working fluid.

Hybrid plants are described so far and combined in various ways to achieve a higher efficiency or to overcome the potential problems related to geofluid characteristics. The following are examples of hybrid plants: direct steam/binary units and flash steam/binary units.

There are three classes of turbine power units used in geothermal plants:

a.) 30 - 40 MWe
b.) 5 - 10 MWe
c.) 0.2 - 0.5 MWe

The power rating for the respective geothermal power cycles are:

Direct steam plant	5 - 40 MWe
Flash steam plant	1 - 35 MWe
Binary plant	0.02 - 0.5 MWe
Hybrid plant	not yet available

The ideal inlet temperature for house heating is about 80°C, but with the application of large radiators with heat pumps or auxiliary boiler, the thermal water with a temperature of only few degree, above the ambient temperature, can be beneficially used.

It uses mostly known technology and straightforward engineering. It is estimated that the installed thermal power, that uses geothermal fluids for direct non-electric application, is over 11 385 MWt, with a total energy production of 31. 10^{12} Kcal. This was achieved with a flow rate of about 84000 kg/s at an average load factor of 36%. These data do not include all uses where the inlet temperature is below 35°C. These also excludes fish farms using warm water at the scale of tens of MWt. This further more ignores most of the heat pumps installations, which have a growing market for space heating and cooling and will become the largest application of geothermal energy systems. Such application of heat pumps reduces the energy consumption by 30% when compared with air source heat pumps and more than 60% when compared with electric or fossil fuel heating and cooling. There are two parameters, which are relevant for the assessment of the geothermal energy for non-electric applications, namely: fluid temperature and fluid flow rate. As regards the temperature range for non-electric geothermal energy use, there are three options:

	Temperature range [°C]
a.) Heating	20 - 80
b.) Drying	80 - 160
c.) Chemical processes	150 - 200

Fluid flow rate is strongly dependent on the respective application capacity. In this respect a rough estimate of the power range for the mentioned applications is the following:

	Power range [MWt]
a.) Heating	10 - 25
b.) Drying	1 - 10
c.) Chemical processes	10 - 100

For each geothermal well temperature, the respective flow rate can be determined and the number of boreholes is estimated in order to meet the specific requirement.

Technologically the geothermal energy application for non-electric use has been demonstrated in a number of countries. The available know-how for different applications has proved its maturity so that, within the defined parameters it can be considered as a demonstrated entity in the technological assessment of NRSE.

Electricity production from geothermal sources cost is around 4 UScent/kWh and for the heat generation the cost is around 2 UScents/kWh [44].

3.3 Biomass Energy Resources

Biomass provide about 14% of the world energy or about 25 millions barrels of oil equivalent per day (Mboe/day) [45,46,47]. It is the most important source of energy in developing countries. In general, it is rather difficult to estimate biomass resources because they are strongly dependent on natural vegetation. Detailed analysis shows that if it is assumed 35 GJ/capita the consumption for developing countries, the land required per capita with biomass yield 2,5 and 10 t/ha/year, will be 1.0 and 0.2 ha/capita, respectively. It is estimated that for biomass use for energy production it would be needed a minimum of 5 t/ha/year yield and it could be used in an area where local energy consumption is substantially below the average for developing countries. Commodities as lighting, water for people, water for live stock and irrigation is of primary interest for the biomass electricity production. As it was shown cooking gas also is estimated for the potential consumer of biomass energy.

In the local resources assessment for biomass energy, main emphasis is given to the land needed and yields of biomass production. The second requirement is strongly dependent on the geographical location and may substantially vary in different locations. As example: in tropical regions the yield is 16 t/ha/ a, while in subtropical the yield is 22 t/ha/ y. So, the parameters to be used in the assessment of biomass local resources are: local demand for land and yield of biomass production.

Biomass energy production can be obtained through different routes for biomass conversion processes. The great versatility of biomass as a feedstock is evident from the range of wet to dry materials, which can be converted into various solid, liquid and gaseous fuels using biological and thermochemical conversion processes. Solid fuels are wood, charcoal, crop and forestry residual, agroindustrial and municipal wastes and briquettes. Biomass derived liquids are mainly ethanol and methanol. Gases are mainly biogases from anaerobic digesters, gasifiers-producing gases which can be used for electricity generation and, possibly, coupled to efficient gas turbines.

There are two main branches of biomass conversion, namely: bioconversion process and thermal process. The bioconversion processes are alcoholic fermentation and anaerobic fermentation. The thermal processes are pyrolisis and combustion of the biomass. In this analysis, priority will be given to those technologies which are demonstrated in industrial scale and could be used as reference for the technological assessment of their maturity. In this respect, bioconversion technology for the ethanol production falls into a cathegory of technologies which are presently commercially available. For the ethanol production there are three options:

	Capacity [l/day]
Micro system	< 200
Mini system	200 - 20 000
Macro system	> 20 000

Figure 12 shows schematic representation of typical biomass power plant.

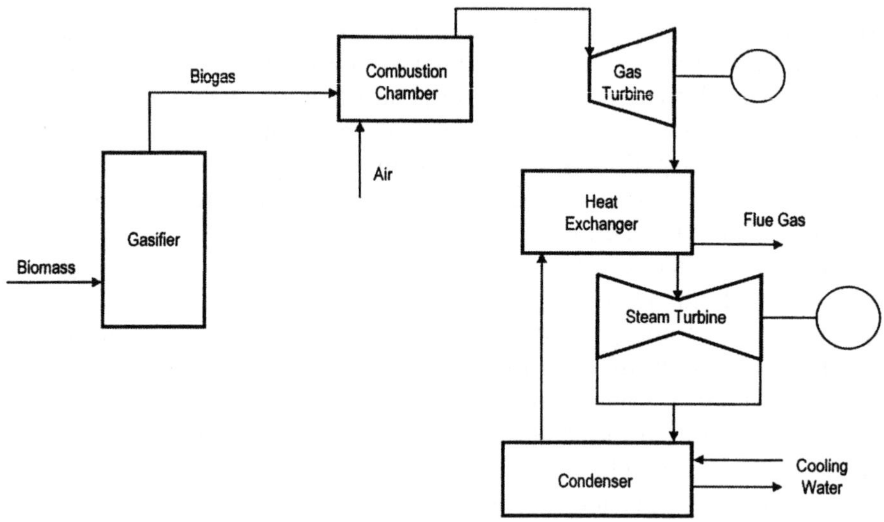

Figure 12. Average size wind turbine.

Among these options, the mini system is ideal from the technical, economical, social and environmental standpoint. It appears to be a highly viable solution to the liquid fuel energy problem, especially for developing countries.

The second option in bioconversion technologies is biogas production. Biogas production is a well-known technology based on anaerobic fermentation. There are a vast number of technical systems used for biogas production. From this analysis it is omitted the biogas production facility based on a primitive technology for individual use.

Pyrolisis and combustion processes are based on high temperature conversion processes with partial or full combustion of biomass fuel. They are used for charcoal or heat production, depending on the process of biomass conversion. The charcoal production plant has not proved its maturity and is yet not available as commercial technology. Cooking stoves are presently commercially available as a combustion technology based product. There are three sizes of the wood fuelled thermal energy plants in cottage, medium and large industry processes and enterprises. The combustion process of biomass is not substantially different from the combustion of any solid fuel. This implies that specially designed boilers are used for the biomass combustion plants. In this respect there are a variety of boiler designs, which may serve this purpose. In particular there is a great interest for the waste incineration boiler as energy source.

It is expected in the medium term, that at a biomass cost of 2 UScents/Gj the electricity produced will reach 10 – 15 UScents/kWh. [48].

3.4 Wind Energy Resources

The world technically exploitable resources are estimated about 300 TWh/year [49,50,51]. They are rather heterogeneous and strongly dependent on the geographical location. It was recognized that the most economical turbines are those with a rating between 1kW and 350 kW. This means that the wind velocity at the potential location of the wind power plant should be a minimum of 6.5 m/s with an availability of 25 - 40%. In this respect, the wind energy to be used for electricity production will depend on the single unit production of the specific size but the total installed capacity may be dependent on local demand or grid capacity. Grid connected turbines in form of wind farms are a forthcoming entity for the wind use for electricity production. In connection with this approach the land occupied by wind farms is to be taken into consideration as local resource of the specific location. It is estimated that 4 - 8 MW could be installed on 1 km^2.

In defining the resource conditions for the wind energy utilization the following parameters are needed:

a.) average wind velocity at the specific location
b.) probability distribution of wind velocity

The wind turbine technology is available in two arrangements of the rotor, in relation to the direction of wind, namely:

a.) Horizontal-axis wind turbine
b.) Vertical-axis wind turbine

The horizontal-axis wind turbine in which the direction of the wind is parallel to the axis has been technologically developed. At present, horizontal-axis wind turbine generators represent approximately 95% of the capacity installed in wind plants. The developed wind turbine plants presently in operation and new turbine could be classified in the following three categories:

		Capacity	Diameter [m]
a.)	Small-size wind turbines	100 kW	< 20
b.)	Medium-size wind turbines	100kW - 1MW	20 – 50
c.)	Large-size wind turbines	> 1MW	> 50

Small sized wind turbine generators are used in a large number of applications. Most of these applications are limited to the supply of isolated dwellings: pumping, desalination, and integration with diesel, storage, integration with other renewable energy sources. In all these applications the storage capacity is an essential factor.

Medium size wind turbine generators are commercially available and grid connected. Figure 13 shows an average size wind turbine installed in the European Community in the period 1986-1992.

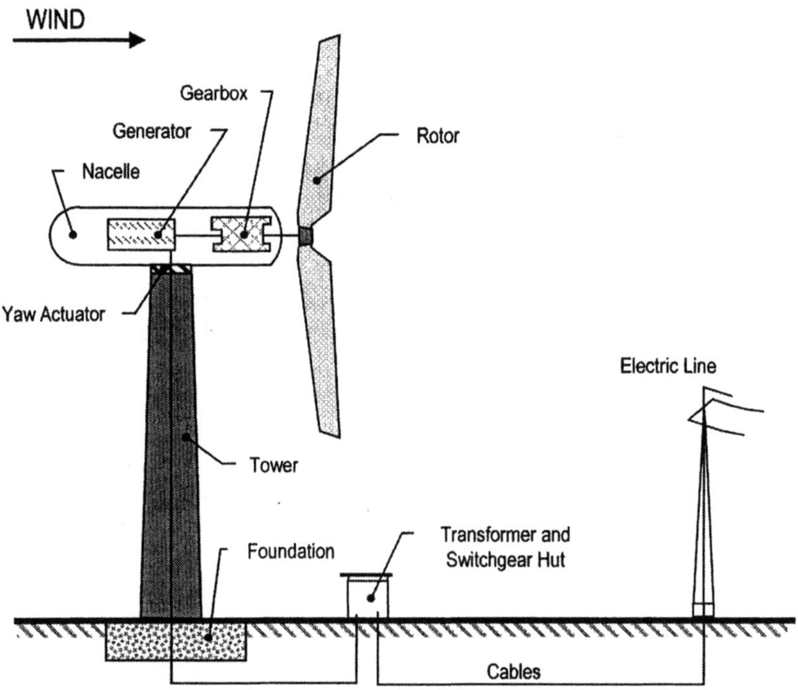

Figure 13. Diagram for selection of respective hydro turbine type.

The configuration of the plant is conventional and the blades are made of glass fibres reinforced plastic or laminated wood bounded by epoxy resin with a power regulation by adjusting the pitch of the blades.
Technology of large-size wind turbine generators is still in the stage of development. Prototypes of the capacity up to 4 MW and rotor diameters up to 100 m are available in some of countries. The obtained results have confirmed the substantial feasibility of large wind generators but have also shown that these machines are still far from being competitive with medium-size machines.

Lowest current electricity cost from wind resources is 6 UScents/kWh [51].

3.5 Hydro Energy Resources

It is estimated that the gross theoretical productivity of hydro energy is about 30 millions GWh/year with an exploitable production of about 13 millions GWh/year [52,53]. The present production of the existing plants is of the order of 2.2 millions GWh/year. The main reason for the large discrepancy between exploitable and present production is the financing capacity of the countries. The average percentage of hydro electric energy production is for developed countries 7% and for developing countries 54%.

In many countries emphasis is given to the utilization of the small-scale hydropower potential. The total capacity of the mini/micro power plant is about 1.5% of the total installed hydropower potential. An equal amount of small-scale plant capacity is currently in the planning stage.

Inventories of suitable hydropower sites have been established for many areas and a vast catalogue of the exploited sites is available. The best sites have already been explored. There is still a lack of knowledge of the potential of small plant sites.

As for any hydraulic turbine, the parameters to be known in the assessment of the specific location are the available hydrostatic pressure and the water flow rate. Hydrostatic height is usually constant for mini/micro hydropower plans. The water flow rate is a very sensitive parameter for small hydropower plants. It varies seasonally and sometime daily. Knowing the respective accumulation, the minimum flow rate can be established. In order to obtain reliable data of the water flow for a small turbine hydropower plants, a rough estimate of the accumulation volume has to be determined and divided by dry season time. Small hydropotential plants can be grouped in three classes:

	Flow rate [m^3/h]	Height [m]
Small hydro power plant	5 - 1000	10 - 300
Mini hydro power plant	0.5 – 50	20 – 50
Micro hydro power plant	0.2 – 3	2 - 50

The hydro energy conversion technology is a mature technology. Long term development of hydro turbines has reached a state where commercially available hydro turbine of all sizes can be purchased in accordance with the respective resource potential. Large hydro turbine plants have been in operation for decades in a number of countries. Demonstration of modern small hydropower plants has been achieved in many countries. These achievements have offered experience and practical training to test the value of the small hydro power development to be used in rural and remote locations.

According to UNIPEDE classification there are three types of hydropower plants, namely

	Power [kW]
Small power plants	< 10 000
Mini power plants	< 2000
Micro power plants	< 500

Technically, there are three main types of hydro turbines, depending on the water flow rate and available height

	Height [m]
Kaplan turbines	2 - 20
Francis turbines	5 - 200
Pelton turbines	50 - 1000

In Kaplan turbines a mechanical momentum is produced by helical blades formed to develop a pressure difference at the front and rear surface of the blades.

The Francis turbine is designed to uptake the water radial flow through fixed blades into rotating blades on the turbine rotor.

The Pelton turbine is designed in the form of one-row double spoon blades exposed to the injected water stream.

The diagram shown in Figure 14 gives the possibility to select the respective hydro turbine type in accordance with the available flow rate and potential height. It also shows the power of the respective turbine to be obtained for the specific parameters of the respective location.

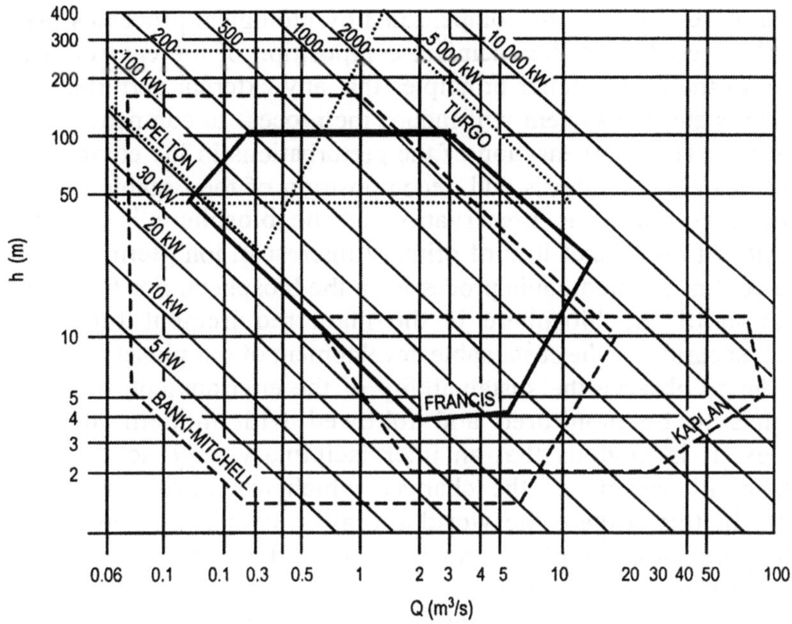

Figure 14. Estimated trend of CO_2 production.

4. ENVIRONMENT CAPACITY FOR THE COMBUSTION PRODUCTS

The sustainability is closely related also to the environment capacity of our planet. It has been shown that the natural processes in the biosphere possess the maximum rate of change. This rate of change exceeds by orders of magnitude the contemporary rates of the parameters defining the antropogenic impact to the environment and by four orders of magnitude, the mean rate of change of the parameters defining the geophysical processes [53]. The concentration and the rate of change of chemicals involved in the biochemical cycles may be characterized by the changes of organic and inorganic carbon. The capacity of biologically active organic and inorganic carbon chemical species in the environment is equal and ten times larger than their annual primary production. Therefore it may be expected that this resource of environment capacity could be considered in the next ten years only if synthesis or decomposition of organic matter is taking place. This means that all life processes will end.

The fluxes of the organic material produced by the synthesis and decomposition processes in the biosphere are within the accuracy of one hundred percent of the antropogenic fluctuation resulting in the environment in the geological time scale. This slow change in the environment in the geological time scale can be compensated by biological processes leading to the biosphere control of the chemical composition of the environment. It is known from the Le Chatelier principles that any external perturbation of the equilibrium state of a system will induce the process to compensate for these perturbations. The compensation of the perturbations in the environment can be obtained by the synthesis and decomposition of the organic processes in the biosphere. Since the preservation of the biosphere is affecting the biodiversity of our planet it is of primary interest in long-term evolution, to have the control of the organic processes in the biosphere. For this reason, the preservation of the biosphere is the main requirement for the global ecological security for the sustainable development of our planet.

In order to observe the sustainability of the environment, the ecological system has to be monitored and followed with modern methods and techniques. It is obvious that an interdisciplinary approach is needed to understand all aspects of the changes which are introduced by human activities. In this respect the world energy system is responsible for the production and emission of a number of chemicals which are proved to have adverse effects on the environment.

The energy use is a major source of emissions. At the same time it is essential to the economic and social development for improved quality of life. There have been recognized several threats as signals for potential hazard to the environment. The emission of air pollutants is usually considered in three groups, namely: carbon dioxide, nitrogen oxide, and sulphur oxides.

The adverse effects of the emission gases are recognized by two processes: the greenhouse effect leading to global warming and depletion of the ozone layer in the stratosphere. The global warming is observed by the increase of the mean earth temperature. It can be noticed that recent changes in concentration of CO_2 in the atmosphere are correlated with the changes in the global temperature. This has led a number of specialists in the field to conclude that the damage is irreversible. Figure 15 shows estimated trend of CO_2 production.

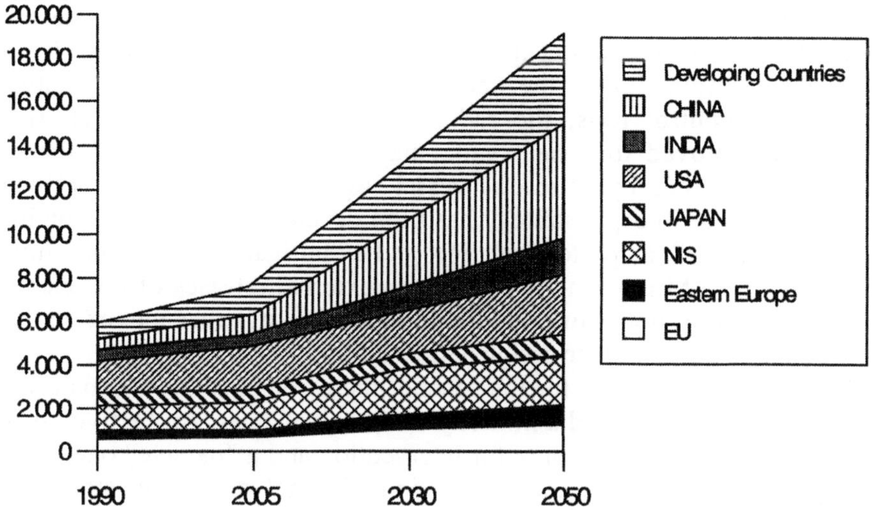

Figure 15. Option for decarbonisation of fossil fuel to electricity and hydrogen.

5. MITIGATION OF NUCLEAR POWER THREAT TO THE ENVIRONMENT

Besides the effects of energy systems on the environment, by the emission of flue gases in the atmosphere due to the combustion of organic fuels in power production plants, there are possibilities which are immanent to some old power plant systems to have hazardous effects on the environment in a very short time with long lasting consequences. The nuclear power plants are very beneficial in light of greenhouse effects because they have no exhaust gases [54]. But, it is known that present nuclear power reactor have the potential possibility to be enormous sources of radioactivity emissions. Besides the effecting the immediate surroundings, these hazardous events may lead to regional and global threat to the environment. The low probability of this kind of events has been only barrier to the disasterous event to spread its consequences to the global environment. Examples recently learned are requiring different approaches to face and master potential hazardous events. Human society is not in position to lean its existence on the man made probability actions without possibility for any correction.

Opponents to nuclear energy outline two points that are crucial for them: the possibility of major radiological releases following to accidents and the heavy inheritage of long lasting radioactive wastes for future generations. Obviously, both these points are very relevant for the sustainability development of this form of energy. Let's discuss separately both. Major accidents may be generated by a reactivity excess (Chernobyl) [55], or by a loss of coolant (Three Miles Island) [56] or by a loss of flowrate or by anticipated transients without the interruption of the nuclear chain. Even if the chain reaction during the accident has been broken with a prompt insertion of control rods, the radioactivity decay residual heat, if not adequately removed to a heat sink, may cause the melting of the core threatening the integrity of reactor vessel. Against these possible accidental chains a "defence in dept" strategy has been developed with three main lines: a "preventive line", a "protective" line, and a "mitigative" line. This strategy worked at Three Mile Island accident with external releases of few curies of radioactivity did not work at Chernobyl due its absence of external containment and many other design deficiencies.

Present reactor designs for the second line of defence have a majority of "active" safety systems and a minority of "passive" ones. An "active" system needs an external energy supply for intervention and a "passive" one is based on physical laws like natural convection, thermal dilatation, stress-strain relations, etc to operate [57]. The present trend of designers is to increase the percentage of passive safety systems to counteract the possible accidental chains, proposing the so called "advanced passive" reactors for a transition period from the first to the second generation of nuclear reactors. This trend is also associated with a preference of a deterministic approach instead of the more scientific probabilistic one, for better gaining the acceptability of common people who often remember the old saying"if it can happen, it will happen".

A design of the second generation of nuclear reactors with 100% passive safety systems already exists. i.e. the MARS (Multi-purpose Advanced Reactor inherently Safe). Figure 16 shows a schematic representation of the MARS reactor [58].

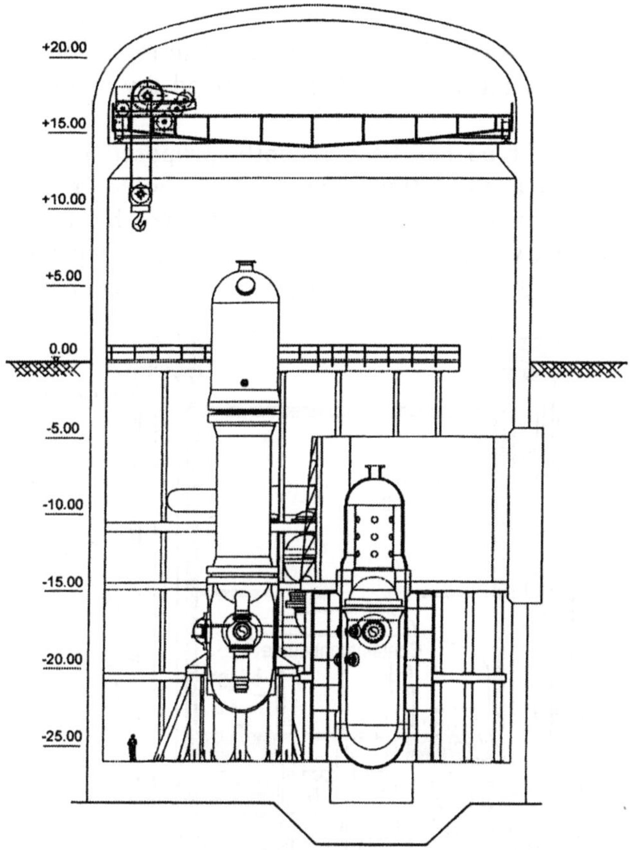

Figure 16. Schematic representation of the MARS reactor.

This reactor has been conceived for small electric networks and for co-generation purposes, to increase the overall efficiency and multiply the possibilities of utilization. It is modular and assembled in small parts that are totally built and controlled, with quality assurance produced in factories. In this way the construction time is shortened, the related cost reduced, and the assembly procedures inverted in order at the end of useful life, to guarantee an easy and total decommissioning of the metal pieces. All these characteristics meet the requirements of sustainability.

The second point of the fear of nuclear opponents, related to the long lasting wastes, is still opened to interesting solutions. It is possible to separate the long-life radionuclides (actinides) from other radioactive fission products. The actinides may be recycled in "ad hoc"reactors and converted in the radionuclides of the (i.e. short life) radionuclides. The radionuclides with long lasting lifetime can be converted into short life isotopes by their mutation through nuclear reactions in high flux nuclear reactors or in a coupled device of subcritical nuclear reactor in which a beam of high energy particles are injected by means of a powerful accelerator. The total amount of long lasting wastes will be substantially reduced so to be conveniently placed in suitable geological formations, which remained dry and intact for million of years in the past.

A device of this type has been recently proposed by Rubbia and co-workers [59], with a subcritical fast reactor cooled by lead in natural convection, feeded by spallation neutrons generated by a beam of protons accelerated till 1 GeV. Figure 17 shows a schematic representation of the conceptual design of this fast neutron operated high power energy amplifier.

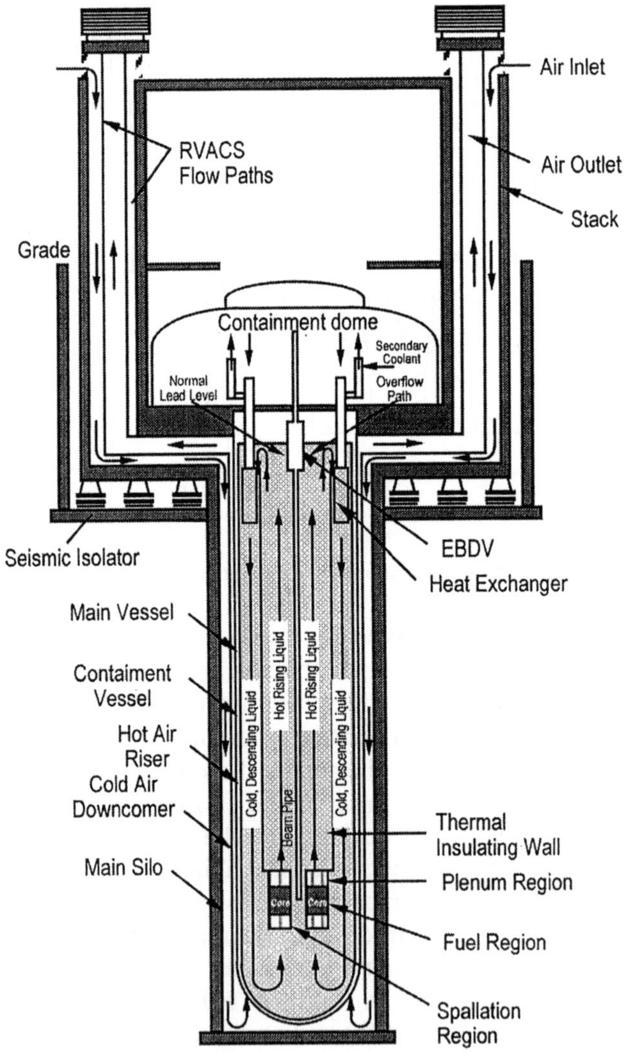

Figure 17. the conceptual design of this fast neutron operated high power energy amplifier.

7. REFERENCES

1. D'Angel, E., Perrella, G., Bianco, R., Energy Efficiency Indicators of Italy, ENEA Centro Ricecche Casaccia, Roma, RT/ERG/96/3, 1996.
2. Angelini, A.M., Actual and Prospective Sources of Electrical Energy, Private Commmunication,1996.
3. Angelini, A.M., A Note on the Role of System Engineering in the Industry of a Changing Society, Technology in Society, Vol. 15, No. 4, pp. 461-466, 1996.
4. Sarkor, A.U., Roragos, S., Sustainable Development of Hydroelectric Energy, Energy, Vol. 20, No. 10, pp. 977-981, 1995.
5. Cafier, G., Conte, G., Rome as a Sustainable City, Agency for a Sustainable Mediterranean Development, 1995.
6. Mitro, B., Lukas, N., Fells, I., European Energy, Energy Policy, Vol. 23, No. 8, pp. 689-701, 1995.
7. Hein, K., Fossil Fuel Utilisation, Third International Conference on Combustion Technologies for a Clean Environment, Lisbon, Portugal, July 1995.
8. El – Sayed, Y., Evans, R., Thermoeconomics and the Design of Heat System, Journal of Engineering for Power, pp. 27-35, January 1970.
9. Sama, D.A., Second Law Insight Analysis Computed with Pinch Analysis a Design Method, Symposium on Exergy Analysis of Energy System, Rome, 1994.
10. EU White Paper, COM (95), 682, An Energy Policy for The European Union, Published by EC, 1995.
11. Furfari, S., The Energy Dimension of EU: Action for a Cleaner Environment, Combustion Technologies for Environment, Lisbon, Portugal, July 1995.
12. Al Gobaisi, D.M.K, The Desalination Manifesto, IDA World Congress on "Desalination and Water Sciences", Vol. 1, pp. 3-27, Nov. 1995.
13. Darwish, M.A., Desalination Processes: A Technical Comparison, IDA World Congress on "Desalination and Water Sciences", Vol. 1, pp. 149-173, Nov. 1995.
14. Mitchelfelder, S., Best Available Technologies for Emission Control in Fossil Fuel Based Power Generation, Fourth International Conference on Combustion for a Clean Air Environment, Lisbon, Portugal, July 1997.
15. Klvana, D., Chuki, J., Guy, C., Kirchneva, J., Catalytic Combustion: New Catalysts for New Technologies, Third International Conference on Combustion Technologies for a Clean Environment, Lisbon, Portugal, July 1995.
16. van Swaaj, W.P.M., Afgan, N.H., Heat and Mass Transfer in Fixed and Fluidized Bed, Hemisphere Publishing Corporation, Washington, 1985.
17. Durao, D.F.G., Heitor, M.V., Moreira A.L.N., On Stabilization of Flame on Multijet Industrial Burners, Experimental Thermal and Fluid Science, Vol. 5, pp. 736-746, 1992.
18. Weinberg, F, Heat Recalculating Burners: Principles, and Some Recent Development, Third International Conference on Combustion Technologies for a Clean Environment, Lisbon, Portugal, July 1995.
19. Carvalho, M.G., Oliviera, P, Semião, V., Three-dimensional Modelling of Glass Furnace, Journal of Institute of Energy, pp. 143-157, 1988.
20. Carvalho, M.G., Coelho, P.J., Heat Transfer in Gas Turbine Combustors, AIAA Journal of Thermophysics and Heat Transfer, Vol. 3, No. 2, pp. 123-11, 1989.
21. Coelho, P.J., Carvalho, M.G., Evaluation of Three-dimensional Mathematical Model of Power Station Boiler, Journal Eng. Gas Turbine, 1986.
22. De Jong, B., Emission Characteristics of the Low-NOx Burners, Third International Conference on Combustion Technologies for a Clean Environment, Lisbon, Portugal, July 1995.

23. Afgan, N., Fundamental Heat and Mass Transfer Research in the Development of the New Heat Exchangers Concept, Journal of Enhanced Heat Transfer, Vol. 2, No. 1-2, pp. 1-16, 1995.
24. Bergles, A., Techniques to Augment Heat Transfer, Heat Transfer Application, ed. by W. Rohsenow, J.Hartnett, E.Ganic, Elsevier Pub. Comp, 1985.
25. Leontiev, A.I., Calculation of the Effectiveness of Cooling Gas Turbine Blades (in Russian), Russian Academy of Sciences-Energetike, No. 6, pp. 85-92, 1993.
26. Goldstein, R.J., Film Cooling Advances in Heat Transfer,Vol. 7, pp. 321-379, Academic Press, 1971.
27. Modern Boiler Design, Private Communication, 1998.
28. Afgan, N. H., Bloch, A. G., Radovanovic, P. M., Zhurovlev, Yu., Gorb, E.I., Boiler Expert System, ICHMT Forum on Expert System and Mathematical Modeling of Energy System, Erlangen, 1991.
29. Afgan, N., Carvalho, M., Coelho, G., P., Concept of Expert System for Boiler Fouling Assessment, Applied Thermal Engineering, Vol. 16, No. 10, pp. 836-844, 1966.
30. Afgan, N. H., He, X. G., Carvalho, M. G., Azevedo, J.L.T., Prototype of Knowledge-based System for Boiler Fouling Assessment at Power Plant Sines, Proceeding of 4th International Conference on Clean Air Technologies, Lisbon, Portugal, 1997.
31. Afgan, N. H., Carvalho, M. G., 1996, An Expert System for Controlling Process in a Power- Station Boiler Furnace, Thermal Engineering, Vol. 43, pp. 514-523.
32. Jamshidi, M., Large –Scale Systems: Modeling, Control, and Fuzzy Logic.
33. Leicester City Council et al., Development of Information System and Computer Models for Energy Management in Urban Environment, EC (XVII/4.1040/95 075G), 1994.
34. Fritz, W., Intelligent Systems and Their Societies, Intelligent System, Home Page, Internet.
35. Berkovski, B. Renewable Energy Sources, UNESCO Contribution to International Cooperation, pp. 845-847, Fueling 21th Century, Ed. A.E. Shendlin, P.Zaleski, Hemisphere Pu. Corp., 1989.
36. Afgan, N., Developing Countries Energy Strategy -Production and Technology Transfer, pp. 43-56, Fuelling 21th Century, Ed. A.E. Shendlin, P.Zaleski, Hemisphere Pu. Corp., 1989.
37. Afgan, N.H., Berkovski, B., Cumo, M., Naso, V., Distant Learning in Energy Engineering Education UNESCO Postgraduate Course, World Congress of Engineering Educators and Industry Leaders, UNESCO, July 1996.
38. Winter, C.J., Sizmann, R.L., Van-Hull, L.L., Solar Power Plants-Fundamentals, Technology, Systems, Economics, Springer-Verlag, Berlin, 1991.
39. Zafran, M., Solar and Health, World Solar Summit, UNESCO, July, 1993.
40. Best, G., Kwschik, Solar Energy and Agriculture, World Solar Summit, UNESCO, July, 1993.
41. Gigopoulos, R, et.al., Solar Energy in Mediteranien, World Solar Summit, UNESCO, July, 1993.
42. Morse, F.H., Photovoltaic Power Systems in Selected IEA Member Countries, Morse Associate, Washington D.C., March 1995.
43. The Intergovermental Panel on Climate Changes, Second Assessment Report, 1996
44. Fridleifsson I.B., Freston D., Geothermal Energy research and Development, World Solar Summit, UNESCO, July, 1993.
45. Dickson, M., Fanelli, M., Geothermal Energy, UNESCO Energy Engineering Series, Edited by B. Berkovski, John Wiley & Sons, 1995.
46. Wereko-Brobby, Ch.Y., Hagen, E.B., Biomass Conversion and Technology, UNESCO Energy Engineering Series, Edited by B. Berkovski, John Wiley & Sons, 1995.
47. Hall, D.O., Biomass Energy, Energy Policy, pp. 711-737, Oct. 1991.
48. Hall, D.O., Rossilo-Calle F., deGroot P., Biomass Energy, Lessons from Case Studies in Developing Countries, Energy policy, pp. 62-73, Jan. 1992.
49. Walker, J.F., Jenkins, N., Wind Energy Technology, UNESCO Energy Engineering Series, Edited by B. Berkovski, John Wiley & Sons, 1995.

50. Sesto, E., Casale, C., Mari, G., Wind Energy – Present Situation and Future Prospects, World Solar Summit, UNESCO, July, 1993.
51. Jiandong, T., Naibo, Z., Xianhuan, H., Huishen, D. Mini Hidropower, UNESCO Energy Engineering Series, Edited by B. Berkovski, John Wiley & Sons, 1995.
52. Cazenave, P. et al., Hydro Electric Power, World Solar Summit, UNESCO, July, 1993.
53. Marchuk, G.I., Kondrotyev, K.Ya., Priorities of Global Ecology, Moscow, Nauka, 1992.
54. Marchetti, C, Nuclear Energy and its Future, International Institute for Applied System Analysis, RP-93-18, 1993.
55. Nenat, J.C., Stavrov, A.I., Sokolski, E, Waight, P.J., Chernobyl Accident: The Consequences in Perspective, Int. Conference on Decade After Chernobyl, Vienna, April, 1996.
56. Three Mile Island Unit # 1, Information, Nuclear Regulatory Commission, Washington, 1995.
57. Cumo, M., Inherent and Passive Nuclear Reactor: Step towards a Second Nuclear Generation, Accademia Nazionale delle Scienze, Vol.113, Serie V, Part 2, 1995.
58. Cumo, M., Energy and Development, 50th Anniversary of the UN and the Italian Contribution toword the realization of the "Earth Chapter", Rome, May 1995.
59. Rubbia, et al., Conceptual Design of a Fast Neutron Operated High Power Energy Amplifier, European Organization for Nuclear Research-CERN/AT/95-44(ET) - Geneva, 29th September, 1995.

CHAPTER 4

SUSTAINABILITY AND ENERGY EDUCATION

1. INTRODUCTION

The contemporary development of the information technology has substantially contributed to the promotion of new means of dissemination information and its multiple use. The set of information itself is a value which may be taken as the entity representing the individual system. The thermodynamic approach to the information has defined it as the state variable which defines the system. It has been shown that the information is the extensive variable of the system and is equal to the logarithm of the number of information. The similarity of the information and entropy of the physical system has led to the use of "negentropy" as the parameter which describes the respective system.

With the development of the information theory and the respective concept of information system, it becomes obvious that the socio-economic system can be described with the approach used in thermodynamics. It was recognized that the socio-economic system goes through changes from one state to other states similarly as in the thermodynamic system. This has lead to the definition of socio-economic system including information as the extensive parameter of system, defined as the negentropy.

The information may have different forms and the knowledge is one of them generated by the scientific and technological research. The knowledge is transferred by the education system. In this respect the scientific and technological development from one side and education system from other side can contribute to the information exchange in any socio-economic system. So, we can consider socio-economic changes in the two processes, namely: the generation of information by the scientific and technological development and the exchange of information by the education process. In this case, to have the harmony in the global system, between the use of its resources and at the same time to take care of the preservation of environment, it is of paramount interest the development of science and technology and the education process in achieving this goal. Also, we can say that the sustainable development of our civilization will require besides using the immediate action to preserve resources and environment capacity for the future generations to devote a substantial attention to the science and technology development as well as the development of the education process. The science and technology development has to be focused to these problems, which will affect the consumption of available material resources and environment capacity.

2. KNOWLEDGE DISSEMINATION

In the development of the sustainability issue there are three fundamental dimensions which have to be taken into consideration in the assessment, namely: knowledge dissemination, science and technology development and exploration of new resources. The knowledge dissemination implies the development of the respective education system which will meet increasing demand for continuos refreshment of the knowledge in order to follow the contemporary development of new technologies.

UNESCO in its background paper for the International Conference on "Environment and Society: Education and Public Awareness for Sustainability" identified some important and interrelated factors contributing to the rising concerns over "sustainability" among the people around the world. These range from the rapid growth and changing distribution of the world population, via the persistence of widespread poverty and environmental degradation to the very notion of "development".

Since the Rio Earth Summit, a new international consensus has emerged concerning the critical role of the education in achieving sustainable development. Principle 36, one of the forty Principles in Agenda 21, concerns" promoting education, public awareness and training". The scope of the Principle 36 is extremely broad from formal education system at all levels, vocational training and development of workforce to education in agriculture and health care, adult education and communication about environment and development. Principle 36 also emphasizes basic education is a prerequisite to environmental and development education, and remains a priority for many countries of the world. Principle 36 is divided into three"programme" areas:

- Reorienting education towards sustainable development
- Increasing public awareness
- Promoting training

"Reorienting education to sustainability requires recognizing that traditional compartments and categories can no longer remain in isolation from each other and that we must work increasingly at the interface of disciplines in order to address the complex problems of today's world. This is true both within education where interdisciplinary is slowly and with difficulty gaining ground, and between the spheres of education, work and leisure as life-long learning emerges as a key concept for planning and developing educational systems. It is also true as concerns the most important boundary of all: that separating those included in education systems from those who are excluded from them...

Ultimately, sustainable development will require an education that not only continues throughout life, but is also as broad as life itself, an education that serves all people, draws upon all domains of knowledge and seeks to integrate learning into all life's major activities...

As concerns sustainable development specifically, it is impossible to predict with reliability what will be the key issues on which people will need information in five, ten, twenty and fifty years. It is predictable, however, that such development will not fit nearly the existing and artificial subdivision of knowledge, which have been in place for more than a century. Hence, understanding and solving problems is likely to require intensified cooperation among scientific fields as well as between the pure sciences and social sciences. Reorienting education to sustainable development will in short, require important, even dramatic changes, in nearly all areas [3]

The life long education is presently recognized as a continuous learning activity to follow the needs of modern society [1] It can be seen as the education system with different levels in accordance with the aim, need and individual perception capability. In particular, the aim of lifelong education is diversified by specific requirements of the professional scope. These programs of lifelong education could be aimed to increase the fundamental knowledge of the specific subject or to the design or technical operation guidance for the respective equipment, system or structure. The need for the lifelong education may come from the individual desire for the increase of his/her competency and also can be the result of the organized University - Industry programs for the upgrading of workforce [2]. Finally, the individual perception capacity may require the additional help in adapting himself/herself to the normal working condition.

3. DISTANT LEARNING

There are different educational schemes which are suitable for an organized approach to the life long education. Distant learning has been introduced in the education system as the associate system to bridge difficulties arising from the lack of space for all those wishing to enter the high education institution. In this respect, the distance learning education methodology has proved advantages, which are of interest to be exposed to the public at large in order to be evaluated through its application in the lifelong education. In particular, it becomes relevant to take the advantages available by the modern communication systems, which are offering large possibilities to convey information to any distance, as desired. If the information to be transferred are adequately organized and structured, the messages will become the knowledge needed to be adapted for the learners.

For this reason, the education and training coupled with the recent development in telematics is opening a new venue for the revolutionary changes in the high education system.

It is also recognized that the present education system is too rigid to accommodate new challenges of the lifelong education system [6]. To assimilate the distant learning education with a modern information technology and its prominence at this time it is important to realize the changes in relationships between high education and society. It is obvious that the present status of the higher education system will no longer enjoy a monopoly on the provision of learning services. Its eminence is blurred by the increasing competency of other institutions penetrating in the education service. Distant learning education with multimedia background has risen to prominence because it has the potential to address the social changes immanent to modern society as an alternative education system. The new distant learning systems will accommodate the time and space of the part time learner, giving the opportunity to those outside metropolitan areas without the prominent education institution to have access to high quality education and to be exposed to outstanding professional specialists in the specific field.

In a number of countries it was proved that the distant learning methodology is a powerful tool for the dissemination of knowledge of different levels. The development of an in-house training and education capacity by the business and other agencies not only signals the rise of importance for the changes of training and education methodology, but also reflects a dissatisfaction with the ability of mainline public institution to provide adaptive, flexible relevant and problem centred approaches to training and education and, moreover, at a reasonable cost.

Particular attention has to be devoted to the postgraduate education, which is socially independent and can be organized without the personal interaction between the teacher and the learners and learners among themselves, as it is necessary for undergraduate education. In the context of the need for efficient and appropriate timing of postgraduate education, the demand and perspective of multimedia distant learning education is of high expectation. Technology oriented postgraduate education programs will be linked into regional, national and international networks and will be a great possibility for the integrated education environment, leading to the democratization of the world education system. The postgraduate engineering education in many contemporary education programs is printed to the continuation of the undergraduate education with emphasis on the fundamental knowledge to be learned or gives the deep knowledge related to specific systems, objects or entities. This subdivision will offer two types of postgraduate education programs, namely, the subject oriented and the object oriented postgraduate education programs.

The modern development of the information technology is offering a challenge for its use in the education process. The multimedia information telematics is one of the new achievements, which is opening a possibility to use the distant learning methodology for postgraduate education. A limited number of qualified teachers for the specific subjects of modern technology is one of the limitations for the knowledge dissemination in the field, needed for the promotion of economic development. The scarcity of competent teachers in specialized disciplines can be adequately subsidized by the respective use of distant learning methodology, which will allow the exposition of the valuable professionals in the field to larger audiences. If consideration is given to the future need for postgraduate education, it becomes obvious that a new lifelong education system is needed to meet the constantly increasing demands for high education in different engineering fields.

4. ENERGY AND ENVIRONMENT ENGINEERING EDUCATION

The energy and environment problems in the contemporary world are the challenging engineering fields requiring the constant increase of our attention in order to meet the increasing demands and also face the potential adverse effects on the environment. In this field, a new strategy is in the sustainable development which compulsory requires the respective efficiency of the systems, increases the global energy system structure use of new and renewable energy sources, meets a new social aspect of the energy and environment. Also it will better understand in deep all consequences which possibly promote global changes in the environment. These requirements could be met only with the respective education systems which will be able to accommodate the needs for an increasing number of specialists with the high level of education, capable to understand in deep all energy and environment problems [4]. It is becoming obvious that the present energy and environment engineering education system is facing difficulties in many countries to meet a forthcoming demand. Even more, in a number of developing countries this problem is open and will require an outside assistance to be overcomed. In this respect, the future energy development strategy has to include the development of the respective energy education system, which will ensure its promotion and practical application. It is of paramount interest to the society to combine its effort in the economic development with the change in the structure of the education system.

The engineering education is closely linked with the economic development [5]. The promotion of new technology advances by the use of information technology is immanent to the further development of the engineering education [6]. In particular, the increase of the engineering knowledge body leading to the larger number of the engineering specializations is only the tip of the iceberg of the problem in engineering education. In this respect, the energy and environment engineering education is of primary interest for the modern society. For this reason, it is needed to look for the new achievements in the education methodological developments. Particular emphasis in this respect has to be given to the postgraduate education in order to ensure the benefits coming from the development in the different fields of science and technology.

Efficiency, reliability, new energy sources and environment are among those issues of energy engineering which are to be recognized as the main driving forces to increase the demand from the energy systems.

The efficiency of the energy systems has been recognized since Sady Carnot has defined the Second Law approach for its assessment. In this respect, it has become important to understand the quality of energy in order to obtain the maximum from a potential source of energy. The exergy analysis has become a powerful tool for the engineering evaluation of the respective energy system. The Second Law efficiency assessment requires the knowledge about principles of the exergy analysis, technical characteristics of the respective system and equipment. The correct efficiency evaluation of the system will require also a social aspect of the energy cost, which include the environment interaction with the respective energy system.

The reliability and safety of the energy systems have become the main issue of the adverse effects, which might have regional, national and global effects. It is a complex issue which requires in deep knowledge about the system its functionality and limitations. This knowledge is required at different stages of the energy system development including design, construction and operation. The risk evaluation is a specific issue, which has to be introduced at each stage of the development. Learning about reliability and safety aspects of the energy system requires a high degree of sophistication of all the subjects to be taught in the normal engineering curriculum. It is known that any malfunction of the energy system leading to the degradation of reliability and the decreasing safety margins of the system is result of complex processes, which are time dependent and interrelated. This implies that in order to gain the knowledge needed for the assessment of reliability and safety of the energy system, the engineering education programs should include time and space dependent processes with the respective mathematical and numerical tools to be used in modern computer systems.

The new and renewable energy sources are based on modern achievements in the physics and chemistry. So, it is immanent to the new energy sources assessment and evaluation to have the respective knowledge in these fields. In particular, it should be emphasized the engineering aspects of the new and renewable energy sources in order to introduce them in the strategy of the energy system development. It is obvious that the energy engineering education has to reflect the needs for the competitive evaluation of these sources of energy. For this reason it is necessary to design the energy education programs taking into consideration the need of society to introduce the new energy sources in the everyday life.

The increased environment concern in the modern society has led to the internationally recognized requirements for the higher attention to the adverse effects of human activity towards its surroundings. In this respect, the energy-producing systems have been called for special attention to introduce the different kinds of measures to prevent consequences leading to the degradation of environment. In view of the impact of energy systems on the environment increased emphasis is given to the assessment of such impacts at the stage of energy programming and the project design in order to take into consideration environment problems caused by the energy production, transport and consumption. Besides the updating traditional curricula in the energy engineering with the environment aspect of the problem, it is needed to include subjects as biology and social sciences in the energy engineering curricula. It has to be emphasized that the engineering education aspect has to be developed and introduced in the energy programs.

5. LIFELONG EDUCATION IN ENERGY ENGINEERING

It is known that the need for upgrading professional capability of the manpower in engineering is the result of the constantly growing demand for the increase of efficiency of technical system and increasing awareness of the environment problems [7, 8]. The organized effort is streamed to meet challenges for the benefits obtained with a new approach to future technical systems and their incorporation in the human environment without adverse effects. One of the lines in this effort is a new lifelong education system. In order to meet this requirement the distant learning education methodology offers the advantages proved to be the cost effective and revealing new features to be enjoyed as the incentive challenge for the learners. In particular, it seems appropriate to focus the attention to advantages which are of common interest for the public at large. In this respect, the distance learning with multimedia is proved to be easily accommodated for the on-job training for the Industry - University co-operation. The education programs could be designed for specific needs and adapted in the on-job environment with limited teaching assistance.

As a milestone of further progress in the process of large-scale extension of the application of scientific and technical achievements, the university and other education institutions should attach importance to the development of postgraduate engineering education closely related to the technology research centers. In this respect, the distant learning with the multimedia education used in an appropriate environment can prove to be cost effective. The multi use of teaching material with a limited number of professionals in many cases has shown to be adaptive to the learner's demand and desire. As most of the postgraduate education programs are closely linked to the specific technology developments, it is of great interest to the design respective courses having in mind the customer interest and local environment. The energy engineering postgraduate courses in this respect has to be aimed to meet the recent strategic demands in the modern development of the global energy system. The variety of options for the specific programs will require a polyvalent approach in the design of the respective postgraduate energy engineering program [9, 10]. It will have to accommodate the different disciplines oriented modules, to be used in the construction of multiple options as required by the individual learners. With this approach, it will be possible to launch the individual education programs, which are to be custom-made to the specific need of the individual learners or their employers. This will lead us to take the full advantage of multimedia environment and if connected with the recent development in telematics may lead to the formation of a global engineering education system. This means that the future learners will have the opportunity to connect themselves to any available education programs in the world network designed as a part of the global system. In this respect, the lifelong engineering education is at the beginning of network formation, which will, similarly to the existing Internet or similar global network, offer the education commodity to any individual wishing to increase his/her competency in the specific field of interest.

6. SHORT COURSES ON SUSTAINABILITY ASSESSMENT OF ENERGY SYSTEMS

Courses are designed to present the state of art in the new and renewable energy sources, clean air technologies and desalination processes and introduce available data for the engineering design of the respective plants. Particular attention will be devoted to the data monitoring and its use in the design of the reliable energy system. This will set a frame for the application of the sustainable assessment method for the individual energy system. Special attention will be devoted to the data collection and presentation for the determination of sustainability indicators for the systems.

The Analysis and Synthesis of Parameters under Information Deficiency Method will be present with the aim to be used as the tool in the Sustainability Assessment procedure.

6.1 Sustainability Assessment of New and Renewable Energy System

The course comprises the following topics:

Programme

Ref	Topic title	Hours
1	Sustainability Concept	2 hours
2	Thermo-economic Analysis of NRES	4 hours
3	Solar Thermal Energy Assessment	4 hours
4	Solar Electric Energy Assessment	4 hours
5	Biomass Energy Assessment	4 hours
6	Wind Energy Assessment	4 hours
7	The Analysis and Synthesis of Parameters under Information Deficiency Method	4 hours
8	Sustainability Assessment of Energy System	4 hours
9	Sustainability Assessment of Energy System - Wokshop	4 hours

Description of the Individual Subjects

1. Sustainability Concept

The first two lectures will be devoted to the introduction of sustainability concept with special reference to its historical background and complexity. Beside giving the energy resource consumption and resource availability, it will be presented overall picture of the present energy state and future strategy reflecting different constrain. Particular attention will be devoted to the potential development of the energy technologies with and its relation to the sustainable development.

2. Thermo-Economic Analysis of New and Renewable Energy Sources

Four lectures of this subject will be devoted to the fundamentals of the exergy analysis and validation of the themo-economic methodology application to the different energy sources evaluation. Attention will be given to the demonstration of the analysis of new and renewable energy systems.

3. Solar Energy Assessment

Within the series of four lectures it will be demonstrated fundamentals of solar energy use, namely: thermal solar, decentralised electric solar conversion and central solar conversion power plant. In particular the emphasis will be given to those solar conversion systems which are commercially available. In this respect, it will be presented special design of solar system and its demonstration to meet respective energy demand. Also, there will be presented the most recent economic evaluation of the available products and operation experience. Application will include PV installation, rural electrification, and water pumping, professional application and grid connection.

4. Biomass Energy Assessment

Four hours lecture will be devoted to the description of the main principles of biomass use for the energy production. Gas fuel and liquid fuel production from biomass will be described. Assessment of environment impacts of biomass production and use will be presented. Cogeneration application of biomass gasifier/gas turbine technologies will be described and evaluated.

5. Wind Energy Assessment

This topic will include assessment of wind energy resources with special reference to the wind characteristics and its evaluation. It will include aerodynamic fundamentals related to the turbine blade design. Configuration of components and operational characteristics will be presented in order to give description of the wind energy system. Economic assessment will include the methodology for economic evaluation of the system and definition of the respective sustainability indicators. In the application part there will be overview of different wind energy systems including: wind-diesel system, wind turbine for water pumping and offshore wind farms.

6. Fuel Cell Assessment

Recently interest in fuel cells has increased sharply and progress towards commercialisation has accelerated. Today practical, competitive fuel cell systems are becoming available and will take a growing share of the markets for power generation equipment and heavy-duty vehicles. Deregulation of the electricity supply industry is changing market. In this respect attention of the Course will be devoted to the sustainability assessment of the fuel cell as the potential future energy technology.

7. The Analysis and Synthesis of Parameters under Information Deficiency Method

The Analysis and Synthesis of Parameters under the Information Deficiency Method represent a modern tool which is used in the evaluation of the cases when deficiency of information exists. This topic will comprise lectures of mathematical fundamental of the method and instruction for the use of the ASPID- 3W code. Special attention will be devoted to the presentation of the procedure to be followed in the application of the method to the evaluation of the energy system.

8. Sustainability Assessment of Energy System

This topic is aimed to present the method of evaluation to be used in the sustainability assessment of energy systems. In this respect it will be defined the set of criteria with corresponding indicators to be used in the energy system assessment. For the selected location it will be demonstrated the method of evaluation and the appropriate selection of the potential option which meets the sustainability criteria taking into consideration non-numerical information.

9. Sustainability Assessment of Energy System – Workshop

The workshop is aimed to exercise the application of the method for the sustainability assessment of the specific new and renewable energy systems. For each student it will be prepared a task with specification of the data to be used in the sustainability assessment of the respective new and renewable energy system.

6.2 Sustainability Assessment of Clean Air Technologies

Programme

Ref.	Topic title	No. of hours
1	Sustainability Concept	2
2	Thermo-economic Analysis of PP	4
3	Clean Air Technologies Development	4
4	Low NO_x Technologies in PP	4
5	New Design of Steam Power Plant	4
6	The Analysis and Synthesis of Parameters under Information Deficiency Method	4

Ref.	Topic title	No. of hours
7	Sustainability Assessment of Energy System	4
8	Sustainability Assessment of Energy System – Workshop	4

Description of the Individual Subjects

1. Sustainability Concept

The first two lectures will be devoted to the introduction of the sustainability concept with special reference to its historical background and complexity. Besides giving the energy resource consumption and resource availability, an overall picture of the present energy state and future strategy reflecting different constrains will be presented. Particular attention will be devoted to the potential development of the energy technologies with and its relation to the sustainable development.

2. Thermo-Economic Analysis of Power Plant

Four lectures of this subject will be devoted to the fundamentals of the exergy analysis and validation of the thermo-economic methodology application to the different power station energy system. Attention will be given to the demonstration of the analysis of to the steam power plant.

3. Clean Air Technologies Development

The series of lectures on clean air technologies will comprise the combustion process and post-combustion processes. The overview of the combustion technologies will be focused on the technology application to minimise costs for the required performance. Particular reference will be given to the solid fuel utilisation in power plant and respective combustion technologies. Post-combustion technologies will include flue gas cleaning technologies, solid precipitation and filtering of flue gas. Special attention will be given to the fuel cell technology and its assessment for small power plant utilisation.

4. Low NO_x Technologies in Power Plants Boiler

There are several technologies which may contribute to the lowering of NO_x concentration in flue gases. It has been proved that respective organisation of combustion process has a potential effect on the NO_x

formation. The re-burning of organic fuel in the power plant boiler is one of the processes that may substantially contribute to the NO_x control. Also, staging the combustion air is favourable for the conversion of fuel bound nitrogen to N_2. In this respect, the method for the assessment of power plant boiler design, with second level of burners and combustion air supply for the NO_x product control, will be developed. In these series of lectures an overview of the available technologies and their demonstration for NO_x control will be presented.

5. New Power Plants

More than fifty percent of the present organic fuelled power plants are coming to a point when some retrofitting is needed in order to meet present limitations imposed by the Kyoto Protocol. Today, the growing concern over global warming is leading to an increased pressure on reducing CO_2 emissions from power stations. Review of the best available technologies for new power generating system will be given. This series of lectures will also outline the future development of emerging technologies. Particular attention will be devoted to the definition of the sustainability indicators for the assessment of power plants.

6. The Analysis and Synthesis of Parameters under Information Deficiency (ASPID) Method

The Analysis and Synthesis of Parameters under Information Deficiency Method represents a modern tool which is used in the evaluation of the cases when deficiency of information exist. This topic will comprise lectures on mathematical fundamentals of the method and instruction for the use of the ASPID-3W code. Special attention will be devoted to the presentation of the procedure to be followed in the application of the method to the evaluation of the energy system.

7. Sustainability Assessment of Energy System

This topic is aimed at presenting the method of evaluation to be used in the sustainability assessment of energy systems. In this respect, the set of criteria with corresponding indicators to be used in the energy system assessment will be defined. For the selected options of the power plant design, the method of evaluation and the appropriate selection of potential options which meets the sustainability criteria taking into consideration non-numerical information will be demonstrated.

8. Sustainability Assessment of Energy System – Workshop

A workshop is aimed at making exercises on the application of the method for the sustainability assessment of the specific new and renewable energy systems. For each student a task will be prepared, with specification of the data to be used in the sustainability assessment of the respective new and renewable energy system.

6.3 Sustainability Assessment of Water Desalination Technologies

Programme

Ref.	Topic title	No. of hours
1	Sustainability Concept	2
2	Sustainability Concept of Water Desalination	2
3	Thermo-economic Analysis of WDP	4
4	Water Desalination Technologies	4
5	Co-generation of Power and Water Desalination Plan	4
6	The Analysis and Synthesis of Parameters under Information Deficiency Method	4
7	Sustainability Assessment of Water Desalination Plant	4
8	Desalination and renewables	4
9	Desalination at third countries	4
10	Sustainability Assessment of Energy System – Workshop	4

Description of the Individual Subjects

1. Sustainability Concept

The first two lectures will be devoted to the introduction of the sustainability concept, with special reference to its historical background and complexity. Besides presenting the energy and water resource consumption and resource availability, an overall picture of the present water and energy state and future strategy reflecting different constrains will be presented. Particular attention will be devoted to the potential development of the water technologies with and its relation to the sustainable development.

2. Sustainability Concept of Water Desalination

These lectures are devoted to the analysis of water resources and role of sea water desalination, management of water resources. Also ethic aspects of the energy use is considered. Special attention will be devoted to the security of the water in the future. The experience of the Gulf States is analysed with special reference to the water security and quality of life.

3. Thermo-Economic Analysis of Water Desalination Plant

Four lectures on this subject will be devoted to the fundamentals of the exergy analysis and validation of the thermo-economic methodology application to the different water desalination systems. Attention will be given to the demonstration of the analysis to the Multi-stage Flashing Water Desalination Plant.

4. Water Desalination Technologies

The series of lectures are devoted to the desalination plant technologies. Attention is focused on the thermodynamic comparison of the Multiple Stage Flashing Process, Multiple Effect Process and Reverse Osmoses process. Design aspect of advanced MSF, MED, RO plants are analysed. For each plant the main parameters are determined in order to define the respective sustainability indicators for the specific design.

5. Co-generation of Power and Water Desalination Plan

Modern water desalination plants are combined with the power production so that co-generation of power and water production has become a common practice in this technology sector. For this a reason, special attention will be devoted to the analysis of co-generation plants and its assessment. In this series of lecture the state of art of co-generation technologies with special reflection to the seasonal operation will be presented. Specially defined sustainability indicators for the co-generation plants will be introduced.

6. The Analysis and Synthesis of Parameters under Information Deficiency (ASPID) Method

The Analysis and Synthesis of Parameters under the Information Deficiency Method represent a modern tool that is used in the evaluation of the cases when deficiency of information exist. This topic will comprise lectures on mathematical fundamentals of the method and instructions for the use of the ASPID- 3W code. Special attention will be devoted to the presentation of the procedure to be followed in the application of the method to the evaluation of the energy system.

7. Sustainability Assessment of Water Desalination System

This topic is aimed at presenting the method of evaluation to be used in the sustainability assessment of water desalination systems. In this respect, the set of criteria with corresponding indicators to be used in the water desalination system assessment will be defined. For the selected options of the desalination plant design, the method of evaluation and the appropriate selection of potential option that meets the sustainability criteria taking into consideration non-numerical information will be demonstrated.

8. Desalination and Renewables

The successful integration of renewable energy technologies such as solar and windpower with desalination plants will be explained, together with practical examples.

9. Desalination at third countries

This topic aims presenting criteria that lead to the installation of desalination plants in third world countries. The economic and social impacts of these type of investments will be presented.

10. Sustainability Assessment of Energy System – Workshop

The workshop is aimed at making exercises for the application of the method for the sustainability assessment of the specific new and water desalination systems. For each student a task will be prepared with specification of the data to be used in the sustainability assessment of the respective water desalination system.

7. REFERENCES

1. Morison, T.R., Beyond Legitimacy: Facing the Future in Distant Learning, International Journal of lifelong Education, 1989.
2. Ramalho, M.F., Co-operation Between Industry and Academia - Distance Learning Approach, International Congress of Engineering Deans and Industry Leaders, June 23-25, 1993, Paris, 1993.
3. Afgan, N., Educational Aspect of Global Energy Strategy International Journal of Global Energy Issues, Vol. 8, No. 1, 1991.
4. Tadmor, Z., Changing Perspective in Engineering Education Proceeding of Engineering Education Meeting The 21 th Century, Neuman Press, 1988.
5. Berkovski Boris, World Strategy Energy Issues, International Journal of Global Energy Issues, Vol. 8, No. 1, 1991.
6. Fazane, Z., The Use of Computers to Support the Learning of New Concepts Innovative methods in Technological Education, Study of Engineering Education, No. 13, 1989.
7. Pudlowski, Z.J., Major Issues and Development Computer-based Instruction Programmes for Continuing Education, International Journal of Continuing Engineering Education, Vol. 1, No. 1, 1990.
8. Afgan, N.H., Durão, D.F.G., Berkovski, B.I., Distant Learning Education for Carrier-Long Training of EngineerInternational Congress of Engineering Deans and Industry Leaders, June 23 -25, Paris, 1993.
9. Energy Engineering International Centre for Heat and Mass Transfer, Doc.1, Doc.2, Doc.3, 1989-1991 .
10. Afgan, N., Cumo, M., Naso, V., Berkovski, B.M., Distant Learning Education In Energy Engineering – UNESCO Postgraduate Course, International Congress of Engineering Education, Rome, June, 1996.

CHAPTER 5

ENERGY SYSTEM ASSESSMENT WITH SUSTAINABILITY INDICATORS

1. INTRODUCTION

The sustainability assessment methodology has to be demonstrated in order to become the tool in the engineering practice. This chapter is devoted to the attempt to evaluate the sustainability of energy systems and show how it can be used in the everyday engineering practice. Obviously, this type of approach has its limitation due to the lack of data for the serious consideration of the system. But it should be anticipated that these excises might serve as guidance for eventual future applications. Also, it should be emphasised that this type of methodology will become an useful tool only if it proves to be justified in the engineering practice.

It should be kept in mind that this example is based on the data collected in the available literature, with insufficient reliability. Also, it should be emphasised that all data are normalised with the total energy production. It is obvious, that this is not an optimal solution and should be reconsidered. In particular this may be important in the assessment of the effect of resources. In the resource evaluation it is adapted to take available total world or local resources as the normalising parameter for the respective indicators.

System selection is also one of the key issues. In order to make analysis simpler, it is adapted to take into consideration the limited geographic area with a rather low level of population index.

The application of sustainability indicators for the assessment of isolated island energy structure is an excise of the use of the presented methodology. In this respect the area and number of habitants define the island with respective economic activity defined by the GNP. It assumed that water has to be produced with the desalination plant.

2. DEFINITION OF SUSTAINABILITY CRITERION

The criteria for the energy system sustainability assessment have to reflect the following aspects, namely: resource aspects, environment aspects, social aspects and economic aspects [1,2]. In this respect, the sustainability assessment of energy system will comprise the evaluation of those parameters, which are the reflection of integral concept of the sustainability. As any other complex system the energy system is defined with constrains which imply its function, technology, geography, property and capacity.

In the definition of sustainability criteria for the energy system the following was taken into consideration in the definition of the criteria [3]:

- to reflect sustainability concept. This will imply that the indicators for respective criteria represent quantities, which have relevancy to the sustainability. This will imply that the energy systems dematerialization in the design may be seen as the introduction of knowledge based systems, use of virtual library, digitalised video, use of on-line diagnostic systems, development of new sensor elements and development of new combustion technologies.
- to be defined with indicators which can be measured and are available as physical parameters representing the data which are possible to obtain with respective quantitative or qualitative form.
- to be based on timely information. The indicator has to be the information, which is relevant to the time to act. This will mean that the energy system and its subsystems have to meet sustainability through every stage of the life cycle. It is known that the energy system works under different conditions in order to meet load change, environment change, social change, etc. It is obvious that there will be different cycles for each of the mentioned time scale processes. In this respect the system has to fulfill its function without failing to meet sustainability requirements.
- to be based on the reliable information. In this respect the indicators have to be the data which you must trust because their totality may be the milestone for the important decision to be made.
- to reflect strategic view. The sustainability is not a quick fix of current problems and it's the way a choosing action today that will cause problems tomorrow. As regards the energy system, it may be interpreted as: mixed energy concept with optimization of local resources; urban and industrial planning with transport optimization; use of the renewable energy sources.
- to give possibility to perform optimisation of the system to minimise energy costs under conditions associated with the available material, financial resources, protection of the environment and government regulations, together with safety, reliability, availability and maintainability of the system.
- to reflect longevity of design. The complex energy system is commonly composed of different subsystems and individual equipment elements. It has been recognized that the life time of the elements and subsystems is not equal. In this respect optimal selection of the life cycle for elements and subsystems may lead to the retrofitting procedure which will reflect

need for the sustainable criterions application. Examples for this criterion can be seen as: modular design of the subsystems, standardization of the elements, lifetime monitoring and assessment, co-ordination of suppliers and buyers.

3. SUSTAINABILITY INDICATORS

Measuring sustainability is a major issue as well as a riving force of the discussion on sustainability development. Developing tools that reliably measure sustainability are a prerequisite for identifying non-sustainable processes informing design-makers of the quality of products and monitoring impacts to the social environment. The multiplicity of indicators and measuring tools being developed in this fast growing field shows the importance of the conceptual and methodological work in this area. The development and selection of indicators require parameters related to the reliability, appropriateness, practicality and limitations of measurement [4].

In order to cope with the complexity of sustainability related issues for different systems the indicators have to reflect the wholeness of the system as well as the interaction of its subsystems. Consequently, indicators have to measure the intensity of the interactions among elements of the system and system and its environment. In this view, there is a need for the indicator sets related to the interaction processes that allow an assessment of the complex relationship of every system and its environment. This will imply that complexity indicators will be defined reflecting links among internal parameters and external parameters of the system. This may be interpreted in the thermodynamic vocabulary as the intensive and extensive parameters of the system.

The effective indicator has to meet characteristics reflecting a problem and criteria to be considered. Its purpose is to show how well system is working. In the case there is a problem that an indicator has to indicate its origin and direction to take to solve the problem. Indicators are strongly dependent on the type of system they monitor.

It is known that any numerical number, semantic expression or mathematical sign is information. Examples of this type of information can be fined in any reading of temperature or other time dependent variables describing state of the system. Information is also parameters used in the design of respective system. Also, the positive or negative signs of the variable are also information.

Collecting information and its processing will convert them in data. So, data represent agglomerated information, which are partially or finally processed. Examples of data can be find as a parameter, which describe evaluated information to be used for the specific purpose. In this respect the average inlet temperature of cooling water in the condenser is data obtained by the averaging procedure adapted for this purpose. Also, heat transfer coefficients used in the design of condenser are the data obtained by the experimental procedure for the heat transfer evaluation.

In order to use the data for the assessment of the respective system, it is necessary to convert them into the indicator. So, the indicator represents the measuring parameter for the comparison between the different states or structure of the system. As an example, the efficiency of the system is the indicator for the quality of energy use in respective system. Also, we can evaluate different structure of the systems by the indicator representing respective entities of the system. In this direction is the assessment of intelligence use in the improvement of the system compatibility with its surroundings, the compatibility being measured by the respective indicators.

In order to quantify criteria for the sustainability assessment of any design of energy system the element indicators are defined to meet this requirement. In this respect, the efficiency of resources use and the technology development are of fundamental importance. It is obvious that the efficiency of energy resource use is a short-term approach, which may give return benefit in the near future. As regard the technology development, the long-term research and development are needed. In some cases it will require respective social adjustment in order to meet requirements of the new energy sources.

3.1 Indicators Definition

Sustainability indicators are by definition parameters, which are used to describe respective criteria aimed to the specific qualification of the system. In this respect the indicators for the energy system are reflecting those properties of the system, which are used to define the system sustainability concept. For the sustainability assessment of energy system the following indicators are used:

- Resource Indicator - RI
- Environment Indicator - EI
- Social Indicator - SI
- Efficiency Indicator - FI

3.1.1 Resource Indicators

The resource indicators [5] for the energy system will comprise three elements including fuel resources, stainless steel resources, cooper resources, and aluminium resources. The indicators reflecting individual element of the resource indicator are defined as the total amount of the respective material resource used in the design of the system divided by the total annual energy production. This means that the following elements will compose the resource indicator.

The fuel indicator will comprise the total organic fuel needed for the annual energy production including fuel consumption for energy production and energy needed for the respective materials production. In this respect, it follows as shown in table 1.

Table 1. Resource indicators.

	Name	Definition	Unit
RI_{fuel}	Fuel Resource Indicator	The amount of fuel consumed in tones divided by energy produced in life-time	kg/kWh
RI_{CS}	Carbon steel Resource Indicator	The amount of carbon steel in tones, used in the construction of the plant divided by energy produced in life-time	kg/kWh
RI_{Coop}	Cooper Resource Indicator	The amount of cooper in tones, used in the construction of the plant divided by energy produced in life-time	kg/kWh
RI_{Al}	Aluminium Resource Indicator	The amount of aluminium in tones, used in the construction of the plant divided by energy produced in life-time	kg/kWh

3.1.2 Environment Indicators

The environment indicator is composed of three elements, namely. CO_2, NO_x SO_2 indicator. Following the same procedure used in the definition of resource indicators, we can adapt that the elements of the environment indicators are shown in table 2.

Table 2. Environment indicator.

	Name	Definition	Unit
EI_{CO2}	Carbon dioxide Environment Indicator	The amount of carbon dioxide in tones produced by the plant divided by energy produced in life-time	kg/kWh
EI_{Nox}	Nitrogen oxide Environment Indicator	The amount of nitrogen oxide in tones produced by the plant divided by energy produced in life-time	kg/kWh
EI_{SO2}	Sulfur dioxide Environment Indicator	The amount of sulfur dioxide in tones produced by the plant divided by energy produced in life-time	kg/kWh
EI_{waste}	Waste Environment Indicator	The amount of waste in tones produced by the plant divided by energy produced in life-time	kg/kWh

3.1.3 Social Indicators

The social indicators reflect the social aspect of the options under consideration [6]. It will comprise the following elements: job indicator, standard indicator and community indicator. The job indicator element represents the number of new jobs to be opened corresponding to the respective option. The standard indicator element reflects the potential increase of the standard of living in the community. The community indicator element takes into consideration the community benefits due to individual option. The social indicator elements are defined as as shown in table 3.

Table 3. Social indicators.

	Name	Definition	Unit
SI_{job}	New Job Indicator	Number of paid hours per kWh produced in life-time	hours/kWh
SI_{Inv}	Capital Indicator	The amount of capital per kWh produced in life-time	USD/kWh
SI_{Div}	Diversity and Vitality Indicator	Number of respective entity per kWh produced in life-time	Number/kWh

3.1.4 Economic Indicators

Economic indicators are based on the elements, including: effectiveness indicator, investment indicator, and energy unit cost indicator [7]. The effectiveness indicator element is defined as the thermodynamic efficiency of the system. It will include the energy efficiency conversion from the energy resources to the final energy. The investment cost indicator is aimed to obtain valorisation of the investment per unit power. The energy unit cost indicator will comprise the cost of the energy per unit kW production.

Following the adapted procedure the economic element indicators are defenied as shown in table 4.

Table 4. Economic indicators.

	Name	Definition	Unit
EcI_{Effic}	Efficiency Economic Indicator	The efficiency of the system divided by the energy production	1/kWh
EcI_{inv}	Capital Investment Indicator	Amount of USD invested in the respective option divided by the energy production in life-time	USD/kWh
EcI_{com}	Community Economic Indicator	Gain of GNP for the community per unit kWh	USD/kWh

4. ENERGY SYSTEM ASSESSMENT

4.1 Energy consumption system

The total demand of energy for the island is composed of the following elements: individual consumption. The number of people living on the respective area and specific energy consumption reflecting standard of living defines individual energy consumption. As the parameters effecting individual energy consumption are: number of people, area and specific consumption.

4.2 Energy Production Systems

It is adapted that the energy structure includes the following energy sources: solar energy [8], wind energy [9], biomass energy [10] and option with electric power production in oil fired thermal power plant.

Solar energy sources are based on photoelectric electric conversion of solar energy based on insulation and respective surface covered by the solar collectors. The solar energy conversion efficiency is used as the parameter reflecting possibility of technology development. Insulation, surface and efficiency are used as the parameters reflecting the possible options to be taken into consideration.

Wind energy source is defined by the average wind velocity and diameter of the windmill. It is assumed that the wind is produced in the wind farm defined by the number of fields.

Biomass energy source is obtained by the assumption of the agriculture waste obtained from the predefined area. It is adapted the range of biomass yield from the minimum to the maximum value with respecting heat values. The efficiency of the thermal plant is taken in the range between minimum and maximum values. Biomass is used as a fuel in thermal power plant with respective efficiency.

Oil resource in thermal plant is used for the cases when indigenous energy source is not sufficient to satisfy local demand. It assumed that important oil is used in order to estimate its requirement.

4.3 Energy Object Structure

The object oriented structure of the Energy object representing Production and Consumption as the subclass of the Energy object is presented in Figure1. Subclasses Production and Consumption are composed of the sub-subclasses

Solar, Wind, Biomass and Oil and Individual, Industry and Desalination with respective attributes. The attributes are defined by the corresponding Minimum and Maximum values defining range of the parameters to be used in the description of any individual option to be taken into consideration.

Minimum and Maximum values for the respective attributes of Energy object. The Energy object is shown on Figure 1.

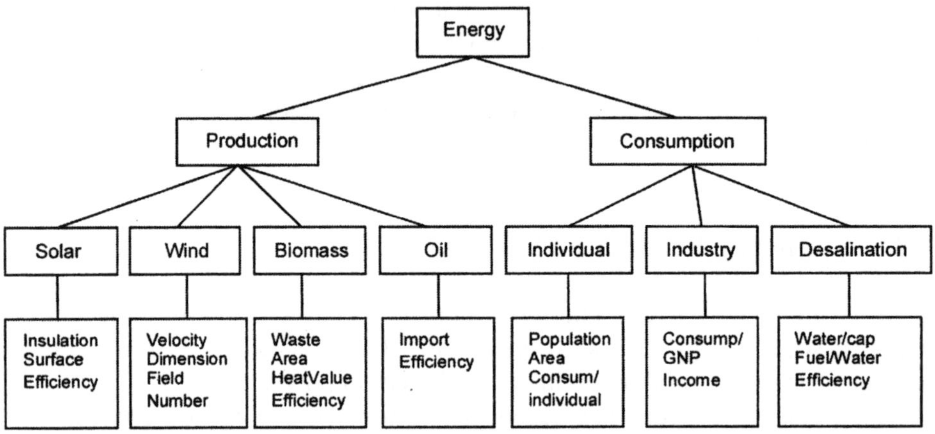

Figure 1. ENERGY object

Table 5 shows the data presenting Minimum and Maximum values for the respective attributes of Energy object.

Table 5. Min and Max values for Energy object.

Energy

				Min.	Max.
Production	Solar	Insulation	W/m^2	80	100
		Surface	m^2	1000	2000
		Efficiency	-	0.1	0.15
	Wind	Velocity	m/s	10	40
		Diameter	m	2	4
		Field	m^2	400	200
		Number	-	20	40
	Biomass	Waste	t/km^2year	100	500
		Surface	km^2	5	10
		Heat Value	kcal/kg	1000	4000
	Oil	Rate	t/year	0	1000
		Efficiency	-	0.25	0.4
Consumption	Individual	Individual	Man/km^2	55	200
		Area	km^2	10	20
		Con/capita	kWh/cap.y	250	2000

Using the presented data in table 5 the following energy production and consumption rate can be obtained for different energy sources. Table 6 presents Minimum and Maximum values of the respective energy sources.

Table 6. Min and Max values for energy sources.

Production		Power		Energy	
		Min	Max	Min	Max
		kW		KWh/year	
	Solar	8	30	$0.0688x10^6$	$0.25x10^6$
	Wind	75	800	$0.645x10^6$	$6.88x10^6$
	Biomass	29	1160	$0.145x10^6$	$5.81x10^6$
	Oil	0		0	$8.8x10^6$
Consumption					
	Individual			$0.125x10^6$	$8x10^6$

For further analysis it will be taken into consideration the number of potential options which are representing specific cases which are in complains respective demand and production alternatives.

4.4 Energy System Options

For the specific geographical site with the defined demand for the electric power several options are taken into consideration. It is aimed to make the assessment of all potential options using respective criteria for the sustainable development. In this respect the following options are taken into consideration:

1. Thermal power plant
2. Solar PV unit
3. Wind power plant
4. Biomass power plant

The nominal power production for the units under consideration is $0.125\ x10^6$ kWh/year. For the respective power production the following energy sources are assumed:

Thermal power plant is CI internal combustion engine with HV = 24 MJ/kg

Solar power plant - average insolation q = 100 W/m^2

Wind power plant -average wind velocity w = 20 m/sec

Biomass fuelled power plant with biomass HV = 15 MJ/kg. The options under consideration are shown in Table 7.

Table 7. Options under consideration.

OPTION UNDER CONSIDERATION				
	Solar	Wind	Biomass	Oil
	kWh/year	KWh/year	kWh/year	kWh/year
Option 1	0,125	0	0	0
Option 2	0	0,125	0	0
Option 3	0	0	0,125	0
Option 4	0	0	0	0,125

4.4.1 Option 1 - Solar Energy

As one of the option to meet energy demand for the energy system under consideration is solar energy use. In this respect, the indicators for the sustainability assessment are defined. As it is specified the energy consumption is

$E_c = 0.125 \text{ x}10^6$ kWh/year

Under solar energy option the PV energy conversion of solar energy is used. The PV panel area for the energy production to meet energy demand is

$$Area = \frac{E_c}{In.Ef.T} = 1.25x10^3 m^2$$

where

Ec- energy consumption
Ef – efficiency of solar energy conversion
T – time period of operation

Under assumption that PV system time period per year is T = 5000 hours/year the power is

$$P_s = 25kW$$

To produce the amount of material for PV panel to be used in PV plant the respective amount of energy is used. Energy needed for the material production of the plant is

$$E_{prod} = PxCxR = 0.112\text{x } 10^6 \text{ MJ}$$

where

P = power produced by PV plant
C_{Al} – = 0.009 t/kW
R – specific energy per unit amount of aluminium = 500 MJ/kg_{Al}
ρ_{Al} - specific density of aluminium = 2.26 g/cm^3

With the assumption that the lifetime of the plant is LT = 10 years the total energy for the production in the lifetime is

$$E_{prod.LT} = E_{con} xLT = 1.25x10^6 \, kWh$$

If it is assumed that the energy production for PV cell panels is

$$E_{prod,tot} = 2xE_{p,Al} = 0.224x10^6 \, MJ$$

If it assumed that the energy for PV panel production is produced in the plant with efficiency $\eta = 0.3$ then the energy resources needed

$$G_{fuel} = E_{tot} xR_{oil} = 0.00934x10^6 .kg_{oil}$$

where

E_{tot} – total energy to be produced
R – specific energy produced by unit of fuel

$$R_{oil} = 0.0417 \, {kg_{oil}}/{MJ}$$

Fuel Resource Indicator is defined as

$$RI_{fuel} = \frac{G_{fuel}}{E_{prod,LT}} = 0.00747.kg_{oil} / kWh$$

where

G_{fuel} – amount of fuel per unit kWh
$E_{prod.LT}$ – total energy consumption per unit kWh

CO_2 Environment Indicator is defined as

$$EI_{CO_2} = \frac{Amount.of.CO_2}{E_{prod.LT}} = \frac{G_{oil} R_{oil}}{E_{prod,LT}} = 0.0186 \frac{kg}{kWh}$$

where

G_{oil} – amount of oil per unit kWh

R_{oil} – amount of CO_2 produced by combustion of oil

$$R_{oil} = C^{kg_{CO2}}\!\!\Big/_{kg_C} \; xD^{kg_C}\!\!\Big/_{kg_{oil}} = 3.11^{kg_{CO2}}\!\!\Big/_{kg_{oil}}$$

Job Social Indicator is defined as

$$SI_{job} = \frac{Number.of.hours}{E_{prod,lT}}$$

It is adapted that the manpower for PV plant is

- 1 operation person
- 2 maintenance person

  Total 3 manpower/year

$$SI_{job} = \frac{N_{man} T[\min/year] T_{LT}[years]}{E_{prod,LT}} = 1.6 \frac{\min}{kWh}$$

Cost Economic Indicator is defined as

$$EcI_{\cos t} = \frac{Cost}{E_{prod,LT}}$$

In the determination of the Economic Indicator a simple procedure is adapted so the Total Cost = Capital Cost + Operation and Maintenance Cost

For this example it is assumed the unit capital cost C = 4400 USD/kW

From this the total cost of the plant is C_C = 25x4400 = 1.1 x 10^5 USD

For this example it is adapted that operation and maintenance cost is C_{O+M} = 2000 USD/year, so that $C_{O+M,tot}$ = 20000 USD

Total Life Cycle Cost $C_{LT} = C_C + C_{O+M}$ = 1.33 x 10^5 USD

$$EcI_{cost} = 10.6 \frac{c}{kWh}$$

It is obvious that the procedure for the Sustainability Indicator calculation can be improved but as this is only example which aimed to present the methodology it is considered that this level of accuracy is sufficient. Solar energy indicators are shown in Table 8.

Table 8. Option 1 – Solar Energy Indicators.

	Fuel Resource Indicator	CO_2 Environment Indicator	Job Social Indicator	Cost Economic Indicator
	kg/kWh	kg/kWh	min/kWh	c/kWh
Option 1	0.00747	0.0186	1.6	10.6

4.4.2 Option 2 - Wind Energy

As the second option under consideration in the exercise is wind energy plant. It is assumed a simple wind plant design for the average wind velocity 10 m/sec. and D = 3 m

Energy Consumption

$$E_{con} = 0.125x10^6\, kWh / year$$

For this exercise it is assumed that the wind plant is the single unit with power

N = 25 kW

Under assumption that the yearly operation time is T = 5000 h/year the total energy production during the year is

$$E_{wind} = 0.125x10^6\, kWh / year$$

In this case the total energy produced in the lifetime is

$$E_{prod,LT} = 1.25x10^6\, kWh$$

It is adapted that the capital cost of the wind plant is assumed C = 600 USD/kW

With the assumption that 50% is material expenses then C_m = 300 USD/kW. Under this conditions the total material expenses are

$C_M = 7500$ USD

It can be used that 1kg of Carbon Steel = 20 USD then the total amount of carbon steel is

$G_{CS} = 0.35$ t

It is known that energy needed for production of 1kg of carbon steel is 200 MJ/kg., so that the total energy needed for the material production for the wind power plant is

$$E_{prod,CS} = 3.5x10^2 x200 = 0.070x10^6 MJ$$

Lifetime energy production is

$$E_{prod,LT} = 1.25x10^6 kWh$$

Total amount of fuel will be

$$G_{fuel} = E_{prod,Cs} xR_{oil} = 0.00292x10^6 kg_{oil}$$

Fuel Resource Indicator is defined as

$$RI_{fuel} = \frac{G_{fuel}}{E_{prod,LT}} = 0.00234 \frac{kg}{kWh}$$

CO_2 Environment Indicator is defined as

$$EI_{CO_2} = \frac{Amount.of.CO_2}{E_{prod,LT}} = \frac{G_{fuel} xR_{oil}}{E_{prod,LT}} = \frac{0.00234x10^6 x3.11}{1.25x10^6} = 0.00582 \frac{kg}{kWh}$$

Job Social Indicator is defined

$$SI_{job} = \frac{Manhours}{E_{prod,LT}}$$

Manpower for wind plant is

- 1 operator

- 1 maintenance
  Total 2 manpower

$$SI_{job} = \frac{N_{man} x T}{E_{prod,LT}} = \frac{2x12x200x60x10}{1.25x10^6} = 2.2 \frac{\min}{kWh}$$

Cost Economic Indicator is

$$EcI_{cost} = \frac{Cost}{E_{prod,LT}}$$

Total Cost = Capital Cost + Operation and Maintenance Cost

C_{cap} = P x K = 25 x 600 = 15000 USD

The operation and maintenance cost is assumed C_{O+M} = 4000 USD/year, so that for the lifetime it will be

$C_{O+M,tot}$ = 40000 USD

Following this simple procedure the total life time cost of the plant C_{LT} = 15000+ 40000 = 55000 USD

$$EcI_{cost} = \frac{55000x10^2}{1.25x10^6} = 4.4 \frac{c}{kWh}$$

Wind energy indicators are shown in Table 9.

Table 9. Option 2 Wind Energy Indicators.

	Fuel Resource Indicator	CO_2 Environment Indicator	Job Social Indicator	Cost Economic Indicator
	kg/kWh	kg/kWh	min/kWh	c/kWh
Option 2	0.00292	0.0052	2.2	4.4

4.4.3 Option 3 - Biomass Energy

The biomass option is based on the power plant fueled by biomass collected from the respective field and gasified in biomass gasifier with gas used in the gas turbine.

Energy consumption

$$E_{con} = 0.125x10^6\, kWh / year$$

Surface area of the biomass is

$$A = \frac{E}{GxHx\eta}$$

where

E - energy consumption
G- gain = 500 t/km^2 year
H-Heating value = 4.652 x 10^3 kWh/t
η - = efficiency = 0.25

$$A = \frac{0.125x10^6}{500x0.25x4.652x10^3} = 0.22.km^2$$

Capital cost defined in the reference[10] with assumption that integrated gasifier is use with gas turbine and steam injection cycle,

C_{cap} = 300 USD/kW

For this exercise it is adapted that operational cost C_{O+M} = 10000 USD/year

Assumed 50% capital cost is the material cost

$$G_{CS} = \frac{C_{cap}\, xP}{2K}$$

where

K = 20 USD/kg_{CS}

$$G_{CS} = \frac{150x25}{2} = 187{,}5.kg_{CS}$$

With 1kg – 200 MJ/kg energy production per kg, the total energy production for the material for biomass power plant is

$$E_{prod,CS} = G_M x K_1 = 3.65x10^4\, MJ$$

The amount of fuel needed for this energy production is

$$G_{fuel} = E_{prod,Cs} x R_{oil} = 0.00152x10^6 kg$$

Fuel Resource Indicator is defined as

$$RI_{fuel} = \frac{G_{fuel}}{E_{prod,LT}} = 0.00122.\frac{kg}{kWh}$$

The lifetime energy production is

$$E_{prod,LT} = E_{con} x T = 1.25x10^6 kWh$$

The total consumption of the biomass will be

$$G_{bio} = \frac{E_{prod.LT}}{H} = 0.30x10^6 kg$$

Total CO_2 production will include two parts, namely: CO_2 production from the biomass fuel and CO_2 production resulting for the energy use for the material production. Following this assumption the total CO_2 production will be

$$CO_{2,tot} = CO_{2,bio} + CO_{2,mat} \quad CO_{2,bio} = G_{bio} x R_{bio} = 0.413.x10^6.kg_{CO2}$$

$$CO_{2,mat} = \frac{E_{prod,CS}}{H} = 0.00152x10^6 kg$$

$$CO_{2,tot} = G_{bio} x R_{bio} + \frac{E_{prod,CS}}{H} x R_{fuel} = 0.415x10^6 kg$$

CO_2 Environement Indicator is defined as

$$EI_{CO_2} = \frac{CO_2}{E_{prod}} = 0.332 \, {kg_{CO2}}/{kWh}$$

$$EI_{CO_2} = \frac{CO_2}{E_{prod}} = 0.332 \, {kg_{CO2}}/{kWh}$$

Job Social Indicator is

In the definition of the social indicator two components are included. First part is operational job and second will include jobs needed for the collection of the biomass.

It is assumed that manpower

Operation 1 manpower
Maintenance 1 manpower
Fuel collection 2 manpower
Total 4 manpower

$$SI_{job} = \frac{Manpower}{E_{prod}} = 4.55.\frac{\text{min}}{kWh}$$

Cost Efficiency Indicator is

$$EcI_{cost} = \frac{Cost}{E_{prod,LT}}$$

Total cost is composed = Capital Cost + Operation and Maintenance Cost
For this exercise it is adapted C_{cap} = 4000 USD/kW, so that the capital cost for the plant

$C_{cap,tot}$ = 25x4000 = 1.0 x 10^5 USD

The operation and maintenance cost is estimated as

C_{O+M} = 10000 USD/year and with T = 10 year the $C_{O+M,tot}$ = 100000 USD

Total Life Cycle Cost

$C_{LC,tot}$ = 2.0x10^5 USD

$$EcI_{cost} = 16\frac{c}{kWh}$$

Biomass energy indicators are shown in Table 10.

Table 10. Option 3 Biomass Energy Indicators.

	Fuel Resource Indicator	CO_2 Environment Indicator	Job Social Indicator	Cost Economic Indicator
	kg/kWh	kg/kWh	Min/kWh	c/kWh
Option 3	0.00152	0.332	4.55	16

4.4.4 Option 4 - Oil Fired Plant

Oil fired power plant option is defined as the single unit Diesel engine with respective axillaries for electric energy production.

Energy consumption

$$E_{con} = 0.125x10^6 \, kWh/_{year}$$

Power needed

$$P = \frac{E_{con}}{T} = 25..kW$$

Adapted operation time per year T = 5000 h/year

For this option it is assumed in accordance with [11] the unit cost of Diesel engine is

C_D = 500 USD/kW

Under this assumption the total cost is

C_{Tot} = 12500 USD

If it is assumed that the material cost is 70% of total cost than

C_{mat} = 10750 USD

From [12] 1kg Carbon Steel $\Rightarrow$ 60 MJ/kg

Total mass of Carbon Steel

It is adapted that 1kg CS $\Rightarrow$ 20 USD so the total amount of material is

G_{CS} = 0.5375 t

Production energy for the respective material is

$E_{prod,CS} = 32250$ MJ

Energy produced in the period of 10 years

$E_{prod,tot} = 1{,}25 \times 10^6 \times 10 = 1.25 \times 10^6$ kWh

The total energy which will be produced in the life of the plant

$E_{tot} = 1.25 \times 10^6$ kWh

Amount of fuel needed in the life time is

$G_{fuel,prod} = E_{prod,tot} \times R_{oil} = 1.25 \times 10^6 \times 0.15 = 0.189 \times 10^6$ kg

$G_{fuel,CS} = E_{prod,CS} \times R_{CS} = 32250 \times 0.0417 = 1344.4$ kg

$G_{fuel,tot} = G_{fuel,prod} + G_{fuel,CS} = 0.1903$ kg

Fuel Resource Indicator is

$$RI_{fuel} = \frac{G_{fuel}}{E_{prod,tot}} = 0.152 \, {kg_{oil}}\!\big/_{kWh}$$

CO_2 Environment Indicator is

$$EI_{CO2} = \frac{Amount..of..CO_2}{E_{prod,LT}} = \frac{G_{fuel} x R_{oil}}{E_{prod,LT}} = 0.47 \, {kg_{CO2}}\!\big/_{kWh}$$

Job Social Indicator is defined as

It is assumed that the following manpower is needed for the plant operation and maintenance.

1 manpower - operation

1 manpower - maintenance

Total 2 manpower

$$SI_{job} = \frac{Manpower}{E_{prod,LT}} = 2.3 \, {\min}\!\big/_{kWh.year}$$

Cost Economic Indicator is determined as

$$EcI = \frac{C_{tot}}{E_{prod,LT}} = 10.6\, {}^{c}\!/_{kWh}$$

The total capital cost is based on the unit cost $C_{cap,unit}$ = 500 USD/kW so that the total capital cost is

C_{cap} = 25x500 = 12500 USD

The operation and maintenance cost is estimated as follows

C_{O+M} = 2x6000x10 = 120000 USD

The total cost is

C_{tot} = 120000 +12500 = 132500 USD

Oil fired plant indicators are shown in Table 11.

Table 11. Option 4 Oil Fired Plant Indicators.

	Fuel Resource Indicator	CO_2 Environment Indicator	Job Social Indicator	Cost Economic Indicator
	kg/kWh	kg/kWh	min/kWh	c/kWh
Option 4	0.152	0.47	2.3	10.6

5. EVALUATION OF THE SUSTAINABILITY INDICATORS

The energy consumption assumption is that the total need of energy in the region under consideration is 0.125 x 10^6 kWh/year and is used only for the domestic use.

Next step in the assessment of the energy system is to determine sustainability indicators. In this respect, in this analysis the following indicators are taken into consideration: Fuel Resource Indicators, CO_2 Environment Indicator, Job Social Indicator and Cost Economic Indicator are shown in Table 12.

Table 12. Indicators for options under consideration.

	Option		RI	EI	SI	EcI
			Kg/kWh	kg/kWh	min/kWh	c/kWh
Solar	1		0,0075	0,0186	1,6	10,6
Wind	2		0,0029	0,0052	2,2	4,4
Biomass	3		0,0015	0,3320	4,5	16
Oil	4		0,1520	0,4700	2,3	10,6
Max			0,1520	0,4700	4,5	16
Min			0,0015	0,0052	1,6	4,4
Max-Min			0,1505	0,4648	2,9	11,6
STDEV			0,0741	0,2317	1,2715	4,7413

5.1 The Assessment Procedure

The assessment procedure is based on the Decision Support System's Shell (DSSS) ASPID-3W [11,12], which is a computer realisation of so called ASPID-method (ASPID [13,14].

The ASPID-technique, which is based on the concept of "randomization of uncertainty", is proposed as a tool for aggregated index construction under shortage ("deficiency") of information. The main ideas of the ASPID-methodology are outlined below and its ability to work with non-numeric (ordinal), non-exact (interval), and non-complete information (NNN-information) is demonstrated. Some potentialities of a flexible interactive decision support system shell ASPID-3W are illustrated by an example of the Energy System options evaluation under uncertainty[15].

Assume that a set $X = \{x^{(j)},\ j = 1,\dots,k\}$ of k *objects* is fixed. In our case *objects* are four *options* of the Energy System ($k = 4$): $x^{(1)}$ - option 1 (Thermal power plant); $x^{(2)}$ - option 2 (Solar PV unit); x^{3} - option 3 (Wind power plant); $x^{(4)}$ - option 4 (Biomass power plant).

Each object $x^{(j)}$ is identified with a vector $x^{(j)} = (x_1^{(j)},\dots,x_m^{(j)})$. A component $x_i^{(j)}$ of the vector $x^{(j)}$ is treated as a value of an initial *attribute* (parameter, variable, characteristic, feature, property, etc.) x_i of the object $x^{(j)}$. In our case initial *attributes* are four *indicators* of the Energy System sustainability ($m = 4$): x_1 - Resource Indicator (RI); x_2 - Environment Indicator (EI); x_3 - Social Indicator (SI); x_4 - Efficiency Indicator (FI).

So, the vector $x^{(j)} = (x_1^{(j)},...,x_m^{(j)})$ may be treated as a value of the attribute-vector $x = (x_1,...,x_m)$. It is supposed that each attribute x_i is necessary and the whole attribute-vector x is sufficient for a fixed *quality* of objects (sustainability - in our case) estimation. In the context of decision-making (i.e., when we must to choose the "best" object-option) the quality under estimation may be identified with the *preferability* of the objects for a fixed decision-maker.

A fixed quality of objects $x^{(j)}$, $j = 1,...,k$, is estimating by many *criteria* $q_1,...,q_m$ each of them being a function of a corresponding attribute: $q_i = q_i(x_i)$, $i = 1,...,m$. A value $q_i^{(j)} = q_i(x_i^{(j)})$ is an estimation of a quality level (degree of preferability) of j-th object from the point of view of i-th *specific (individual) criterion* (*index, indicator*). Without the loss in generality it may be supposed that all specific criteria are normalized, i.e., any criterion q_i meets the inequality $0 \le q_i \le 1$. As this *normalization* takes place, the minimal value $q_i^{(j)} = 0$ of i-th criterion is correlated with an object $x^{(j)}$, which has minimal degree of preferability (from the point of view of i-th criterion), and maximal value $q_i^{(l)} = 1$ - with an object $x^{(l)}$ which has maximal degree of preferability (from the same point of view). So, every object $x^{(j)} \in X$ gets a *multicriteria estimation* $q^{(j)} = (q_1^{(j)},...,q_m^{(j)})$, $0 \le q_i^{(j)} \le 1$, which may be treated as a value of the criteria-vector $q = (q_1,...,q_m)$.

By way of example let us consider exponential normalization function $q_i = q_i(x_i) = q_i(x_i;\theta)$ (θ - a parameter of convexity) which falls into two types. If preferability of any object is increasing (from the point of view of i-th criterion) with increasing of the argument x_i then we shall use increasing normalization function determined by the formula

$$q_i(x_i;\theta) = \begin{cases} 0 & \text{if} \quad x_i \le MIN_i, \\ \left(\dfrac{x_i - MIN_i}{MAX_i - MIN_i} \right)^{\theta} & \text{if } MIN_i < x_i \le MAX_i, \\ 1 & \text{if} \quad x_i > MAX_i. \end{cases} \tag{1}$$

If preferability of any object is decreasing (from the point of view of i-th criterion) with increasing of the argument x_i then we shall use the decreasing normalization function determined by the formula

$$q_i(x_i;\theta)=\begin{cases} 1 & if \quad x_i \le MIN_i, \\ \left(\dfrac{MAX_i - x_i}{MAX_i - MIN_i}\right)^{\theta} & if \; MIN_i < x_i \le MAX_i, \\ 0 & if \quad x_i > MAX_i. \end{cases} \qquad (2)$$

The value MIN_i (MAX_i) in (2) may be chosen by the simplest way as the minimal (maximal) value $MIN_i = \min_j\{x_i^{(j)}, j=1,...,k\}$ ($MAX_i = \max_j\{x_i^{(j)}, j=1,...,k\}$) from the set $\{x_i^{(j)}, j=1,...,k\}$. In actual practice, the most popular normalization function is the simplest one - piecewise linear function $q_i(x_i;\theta)$, $\theta=1$. This simplest and popular linear normalization gives in our case values of normalized indicators (individual criteria, specific indices, etc.) (see Table 13): $q_1 = q_1(x_1) = q_1(RI)$, $q_2 = q_2(x_2) = q_2(EI)$, $q_3 = q_3(x_3) = q_3(SI)$, $q_4 = q_4(x_4) = q_4(EcI)$. So, j-th row in Table 13 may be treated as a multicriteria estimation $q^{(j)} = (q_1^{(j)},...,q_4^{(j)})$ of j-th option, $j=1,...,4$. Values normalized indicators for sustainability of energy system options are shown in Table 13.

Table 13. Values of normalized indicators (specific indices) of Sustainability of the Energy System's options.

Option	RI	EI	SI	EcI
Option 1	0.7342	0.8394	0.0000	1.0000
Option 2	0.9662	1.0000	1.0000	0.4516
Option 3	1.0000	0.0000	0.3333	0.0000
Option 4	0.0000	0.3610	0.8333	0.4839

The values $q_i^{(j)}$, $i=1,\dots,m$, $j=1,...,k$, of specific indices $q_1,...,q_m$ for all objects from the set X being fixed we can try to compare objects *general preferability* with help of component-wise preference relation $\succ$ determined on the $X = \{x^{(j)}, j=1,...,k\}$ by the condition

$$\forall x^{(j)}, x^{(l)} \llcorner X \; (x^{(j)} \succ x^{(l)}) \; \int \; ((\forall i \; q_i^{(j)} \; ? \; q_i^{(l)}) \; \& \; (\exists s\!: q_s^{(J)} > q_s^{(l)})). \quad (3)$$

In plain words, object $x^{(j)}$ is more preferable "in general" than object $x^{(l)}$ ($x^{(j)} \succ x^{(l)}$) if and only if object $x^{(l)}$ is not more preferable than $x^{(j)}$ from the point of view of each specific criterion q_i ($q_i^{(j)} \; ? \; q_i^{(l)}, i = 1, \ldots, m$) and there exists a specific criterion q_s such that $x^{(j)}$ is more preferable than $x^{(l)}$ from the point of view of the criterion ($q_s^{(j)} > q_s^{(l)}$).

Two objects $x^{(j)}, x^{(l)} \llcorner X$ are named *incomparable* if they meet the condition

$$(\exists r\!: q_r^{(j)} > q_r^{(l)}) \; \& \; (\exists s\!: q_s^{(j)} < q_s^{(l)}). \quad (4)$$

For example, multicriteria estimations $q^{(1)} = (q_1^{(1)}, \ldots, q_4^{(1)}) \cup (0.73, 0.84, 0.00, 1.00)$, $q^{(2)} = (q_1^{(2)}, \ldots, q_4^{(2)}) \cup (0.87, 1.00, 1.00, 0.45)$ of options 1,2 sustainability are incomparable as Option 2 is "better" (more preferable) than Option 1 by the criteria q_1, q_2, q_3, but Option 1 is "better" than Option 2 by the criterion q_4 (see Table 9).

The set of such pairs of "incomparable" objects constitutes the major portion of the set of all possible pairs of objects from the set X. So, immediate multicriteria comparison of complex objects meets *the problem of incomparability* rather often. Really, assume that two objects are independently taken by chance from an infinite set of all possible objects determined by criteria-vectors $q = (q_1, \ldots, q_m)$, $0 \le q_i \le 1$. The probability of these two objects incomparability is equal to

$$P(m) = 1 - \frac{1}{2^{m-1}} \quad (5)$$

The problem of objects multicriteria incomparability may be decided by synthesis (aggregation, convolution, etc.) of specific criteria (indices) $q_1, \ldots, q_m$ into one *general criterion* (index) Q determined by a scalar-valued synthesizing (aggregative) function ("convolution") $Q = Q(q) = Q(q_1, \ldots, q_m)$ which meets the condition of monotonicity

$((x^{(j)} \succ x^{(l)})\ ?\ (Q(q^{(j)})\ ?\ Q(q^{(l)})))$ and fills the requirements: (1) $Q(0,...,0)=0$; (2) $Q(1,...,1)=1$. Inequality $Q(q^{(j)}) > Q(q^{(l)})$ means that j-th object is more preferable (from the point of view of general criterion Q) than l-th object. Now all objects are comparable (by the general criterion Q) as there are only three alternatives for any pair of objects $x^{(j)}$, $x^{(l)}$: $Q(q^{(j)}) > Q(q^{(l)})$; $Q(q^{(j)}) < Q(q^{(l)})$; $Q(q^{(j)}) = Q(q^{(l)})$.

There are a lot of possible synthesizing functions Q but usually the simplest modification of the generalized weighted mean is in use, namely *additive aggregative function* (*weighted arithmetical mean*)

$$Q_+(q;w) = \sum_{i=1}^{m} w_i q_i \quad . \tag{6}$$

In this formula $w=(w_1,\ldots,w_m)$, $w_i\ ?\ 0$, $w_1+\ldots+w_m=1$, is a vector of *weight-coefficients* $w_1,\ldots,w_m$ (*weight-vector*). A weight-coefficient w_i is a *measure of relative significance* of the corresponding specific criterion q_i for aggregative estimation $Q(q^{(j)})$ of general preferability of an object $x^{(j)} \llcorner X$.

Weighted arithmetical mean is the most popular type of synthesizing functions. And there are some reasons for such popularity of the additive aggregative function $Q_+(q;w)$. This function is the most simple and easy interpretable synthesizing function it is a quite natural form of specific criteria aggregation for majority of real decision-makers [17].

To summarize, we say that outlined *general indices method* (GIN-method) may be represented as a series of the following steps:

(1) initial attributes $x_1,\ldots,x_m$ for complex objects quality (preferability) estimation are fixed;

(2) specific criteria (indices) $q_1,\ldots,q_m$ for the fixed quality evaluation are chosen and represented in the form $q_i(x_i)$, $0 \le q_i \le 1$, $i=1,\ldots,m$, where $q_i(x_i)$ is a normalization function;

(3) an aggregative function $Q(q;w) \llcorner [0,1]$ of multicriteria estimation $q=(q_1,\ldots,q_m)$ with a vectorial parameter $w=(w_1,\ldots,w_m)$, $w_i\ ?\ 0$, $w_1+\ldots+w_m=1$, is selected for specific indices (specific criteria)

$q_1,...,q_m$ convolution into one general index (general criterion) $Q = Q(q;w)$;

(4) an unique value of the vectorial parameter $w = (w_1,...,w_m)$ is chosen from the set $W = \{w = (w_1,...,w_m): w_i\ ?\ 0, w_1+...+w_m = 1\} \{\ E^m$ of all possible values of the parameter with consideration for the interpretation: each component w_i ("weight-coefficient", "weight") of the vector w is a measure of i-th specific index q_i relative significance for general estimation $Q(q^{(j)};w)$ of preferability of an object $x^{(j)}$.

To distinguish between indices (criteria) $q_1,...,q_m$ (which estimate objects quality (preferability) from different "points of view") and an index (criterion) Q (which evaluates objects quality in general) we prefer to use the pair of antonyms "specific-general" but other authors have a wide selection to pick from the large set of antonyms pairs: local-global, particular-common, individual-collective, isolated-aggregated, analytic-synthetic, differential-integral, etc.

The additive form $Q_+(q;w)$ for the aggregative function being fixed we can finish the process of specific indices $q_1,...,q_m$ synthesizing into general index Q by choice of an element from the set (simplex) $W = \{w = (w_1,...,w_m): w_i\ ?\ 0, w_1+...+w_m = 1\}$ of all possible weight-vectors. But it is well known that the most subtle and delicate stage in GIN-method is just the stage of estimation of "weights" $w_1,...,w_m$ because of usual shortage of an information about exact numerical values of the weight-coefficients. As a rule, we have only *non-numeric* (*ordinal*) *information*, which can be represented by a system

$$OI = \{w_i > w_j, w_r = w_s, ... i,j,r,s,... \llcorner\ \{1,...,m\}\} \tag{7}$$

of equalities and inequalities, and/or *non-exact* (*interval*) *information* which can be represented by a system

$$II = \{0 \le a_i \le w_i \le b_i \le 1,\ i \llcorner\ \{1,...,m\}\} \tag{8}$$

of inequalities which determine intervals $[a_i, b_i]$ for the weight-coefficients possible values. It may be so that some weight-coefficients do not enter in these systems OI, II of equalities and inequalities. In such case we can say that information $I = OI \mid \; II$ is *non-complete*. So, we can speak about non-numeric non-exact and non-complete information (*NNN-information*).

Let us suppose that measurement of weight-coefficients is accurate to within a step $h = \frac{1}{n}$, n - a positive integer. In this case the infinite set W of all possible weight-vectors may be approximated by a finite set (discrete simplex)

$$W(m,n) = \{w^{(t)} = (w_1^{(t)},\dots,w_m^{(t)}) : w_i^{(t)} \llcorner \; \{0, 1/n, 2/n, \dots, 1\}, w_1^{(t)} + \dots + w_m^{(t)} = 1\} \tag{9}$$

of all possible weight-vectors with discrete components. The number of elements of the set $W(m,n)$ may be counted by the formula

$$N(m,n) = \binom{n+m-1}{n} = \binom{n+m-1}{m-1} = \frac{(n+m-1)!}{n!(m-1)!} \tag{10}$$

The number $N(m,n)$ increases swiftly with numbers m,n increasing: $N(5,20) = 10626$, $N(10,40) = 2054455634$ - for instance. In our case it is supposed that the measurement of weight-coefficients is accurate to within the step h = 1/n = 1/20 = 0.05. So, we have $N(4,20) = 1771$ possible weight-vectors $w^{(t)} = (w_1^{(t)},\dots,w_4^{(t)})$, $t = 1,\dots,N(4,20) = 1771$.

Now a NNN-information $I = OI \mid \; II$ may be used for reducing of the set $W(m,n)$ of all possible vectors $w^{(t)} = (w_1^{(t)},\dots,w_m^{(t)})$ with discrete components to a set $W(I;m,n) = \{w^{(s)}, s = 1,\dots,N(I;m,n) \le N(m,n)\} \mid \; W(m,n)$ of all admissible weight-vectors, i.e., weight-vectors which meet the requirements implied by the additional information $I = OI \mid \; II$. Let an expert supposes, that the weight-coefficients $w_1,\dots,w_4$ are arranged in the order $I = \{w_1 > w_2 > w_3 > w_4\}$. In this case we have only $N(I;4,20) = 47$ admissible weight-vectors $w^{(t)} = (w_1^{(t)},\dots,w_4^{(t)})$, $t = 1,\dots,N(I;4,20) = 47$.

A set of all admissible weight-vectors $W(I;m,n)$ being fixed, we have the set $Q^+(I;m,n) = \{Q_+^{(s)}(q) = Q_+(q;w^{(s)}), s = 1,\dots,N(I;m,n)\}$ of all admissible aggregative functions. A function $Q_+^{(s)}(q)$ from the set $Q^+(I;m,n)$ determines for an object $x^{(j)}$ corresponding general index value $Q_+^{(s)}(q^{(j)})$, $s = 1,\dots,N(I;m,n)$.

The average general estimation $\overline{Q}_+(q;I)$ may be calculated by the formula

$$\overline{Q}_+(q;I) = \frac{1}{N(I;m,n)} \sum_{s=1}^{N(I;m,n)} Q_+^{(s)}(q) = \frac{1}{N(I;m,n)} \sum_{s=1}^{N(I;m,n)} Q_+(q;w^{(s)}), \quad (11)$$

where $w^{(s)} \lfloor W(I;m,n)$. This function $\overline{Q}_+(q;I)$ implicitly contains NNN-information I. So, we can treat the values $\overline{Q}_+(q^{(1)};I),\ldots,\overline{Q}_+(q^{(k)};I)$ as the desired *average general estimations* (of the objects $x^{(1)},\ldots,x^{(k)}$ preferability (quality)) which are derived under deficiency of numerical information about weight-coefficients $w_1,\ldots,w_m$.

The exactness of an average general estimation $\overline{Q}_+(q^{(j)};I)$ of j-th object's preferability may be measured by the standard deviation

$$S_j = S(q^{(j)};I) = \sqrt{\frac{1}{N(I;m,n)} \sum_{s=1}^{N(I;m,n)} [Q_+^{(s)}(q^{(j)}) - \overline{Q}_+(q^{(j)})]^2}\,. \quad (12)$$

The reliability of preference relation between objects $x^{(j)},x^{(l)} \lfloor X$ (for the sake of definiteness, we assume that $\overline{Q}_+(q^{(j)};I) > \overline{Q}_+(q^{(l)};I)$) for corresponding average general estimations) may be measured with help of the ratio

$$P(j,l;I) = \frac{|\{s: Q_+^{(s)}(q^{(j)}) > Q_+^{(s)}(q^{(l)})\}|}{N(I;m,n)}, \quad (13)$$

where $|\{s:\ \ldots\ \}|$ is the number of elements of the finite set $\{s:\ \ldots\ \} | \ \{1,\ldots,N(I;m,n)\}$.

Uncertain determination of the discrete weight-coefficients may be interpreted in probabilistic terms. Really, uncertain choice of a weight-vector $w = (w_1,\ldots,w_m)$ from the set $W(I;m,n)$ may be modeled by a random choice, which is specified by uniform distribution on the set. Such *randomization of uncertainty* gives us random weight-coefficients $\widetilde{w}_1(I),\ldots,\widetilde{w}_m(I)$. Mathematical expectation $\overline{w}_i(I) = M\widetilde{w}_i(I)$ of i-th random weight-coefficient $\widetilde{w}_i(I)$ may be used as a numerical estimation of i-th particular criterion significance and exactness of this estimation may be measured by standard deviation $s_i(I) = \sqrt{D\widetilde{w}_i(I)}$ of the random variable $\widetilde{w}_i(I)$. As a vector $\overline{w}(I) = (\overline{w}_1(I),\ldots,\overline{w}_m(I))$, $\overline{w}_i(I) \ ? \ 0$,

$\overline{w}_1(I)+\ldots+\overline{w}_m(I)=1$, of numerical estimations of weight-coefficients is determined on the base of information I, this vector may be treated as a result of arithmetization of NNN-information I and may be interpreted as "*Numeric Image of Nonnumeric Information*", or as "*Quantitative Image of Qualitative Information*".

The reliability of preference relation between weight-coefficients $\overline{w}_r(I)$, $\overline{w}_s(I)$ (for the sake of definiteness, we assume that $\overline{w}_r(I)>\overline{w}_s(I)$) may be measured with help of the ratio

$$p(r,s;I)=\frac{|\{t:w_r^{(t)}>w_s^{(t)}\}|}{N(I;m,n)}, \tag{14}$$

where $|\{t:...\}|$ is the number of elements of the finite set $\{t:...\}|\ \{1,...,N(I;m,n)\}$. From the probabilistic point of view this value $p(r,s;I)$ is probability of the stochastic inequality $\widetilde{w}_r(I)>\widetilde{w}_s(I)$.

Random weight-coefficients $\widetilde{w}_1(I),...,\widetilde{w}_m(I)$ being substituted in additive aggregative function (6), we get *randomized general index*

$$\widetilde{Q}_+(q;I)=Q_+(q;\widetilde{w}(I))=\sum_{i=1}^{m}q_i\,\widetilde{w}_i(I)\ , \tag{15}$$

which is a random variable distributed on the interval $[0,1]$. Now, an average general estimation $\overline{Q}_+(q;I)$ calculated by the formula (11) may be treated as mathematical expectation $\overline{Q}_+(q;I)=M\widetilde{Q}_+(q;I)$ of the corresponding randomized general index $\widetilde{Q}_+(q;I)$. In the case formula (12) gives standard deviation $S(q^{(j)};I)=\sqrt{D\widetilde{Q}_+(q^{(j)};I)}$ of the randomized general index ($D\widetilde{Q}_+(q^{(j)};I)$ - variance of the random variable $\widetilde{Q}_+(q;I)$). And, at last, the value $P(j,l;I)$, determined by formula (13), is the probability of the stochastic inequality $\widetilde{Q}_+(q^{(j)})>\widetilde{Q}_+(q^{(l)})$. In other words, $P(j,l;I)$ is the probability that random general estimation $\widetilde{Q}_+(q^{(j)};I)$ dominates random general estimation $\widetilde{Q}_+(q^{(l)};I)$.

To sum up the outlined method of randomized general indices, we can present next consequence of steps of ASPID-3W program.

An user starts to work with the DSSS ASPID-3W by fixing a list of initial parameters $x_1,\ldots,x_m$ and a list of objects under investigation. Then he must enter a matrix $(x_i^{(j)})$, $i=1,\dots,m$, $j=1,\dots,k$, where element $x_i^{(j)}$ is a value of i-th parameter for j-th object. The following step consists in normalization functions $q_1(x_1),\dots,q_m(x_m)$ formation. For every initial parameter x_i the user must: (1) to fix two values $MIN(i)$, $MAX(i)$; (2) to indicate is the function $q_i(x_i)$ decreasing or increasing with argument x_i increasing; (3) to choose the exponent's value θ in the formulas (1), (2). The functions $q_1(x_1),\dots,q_m(x_m)$ formation process being finished, DSSS ASPID-3W computes a matrix $(q_i^{(j)})$, $i=1,\dots,m$, $j=1,\dots,k$, where an element $q_i^{(j)}$ is a value of i-th specific criterion for j-th object.

Now the user must to fix a step $h=\frac{1}{n}$ (it is supposed that measurement of weight-coefficients is accurate to within h) and insert interval information II about weight-coefficients $w_1,\dots,w_m$ in the form of inequalities: $Min(i)\le w_i\le Max(i)$, $i=1,\dots,m$. Here by default: $Min(i)=0$, $Max(i)=1$. The DSSS recalculates limits for weights' values taking in consideration discrete nature of weight-coefficients and the fact of weights normalization ($w_1+\dots+w_m=1$).

Then the user inserts ordinal information OI in the form of pair-wise inequality and equality relations: $w_i > w_j$, $w_r = w_s$, ..., $i,j,r,s,\ldots \in \{1,\dots,m\}$. After checking of contradictions in ordinal information OI ASPID-3W formats on the set $\{w_1,\dots,w_m\}$ an order relation which takes in consideration all implications of initial pair-wise inequality and equality relations.

All preceding procedures being done, the user, in fact, transforms the DSSS into a DSS: the DSS' shell is filled with information. Really, the user has inserted into ASPID-3W data: (1) a list of initial parameters; (2) a list of objects under investigation; (3) a matrix $(x_i^{(j)})$; (4) a type of each function $q_i(x_i)$ (is it decreasing or increasing function); (5) a value of the parameter θ; (6) limits $MIN(i), MAX(i)$; (7) a step $h=\frac{1}{n}$; (8) an interval information II about weight coefficients; (9) an ordinal information OI about weight coefficients.

Now, the formed DSS ASPID-3W generates all possible weight-vectors $w^{(t)}=(w_1^{(t)},\dots,w_m^{(t)})$, $w_i^{(t)}\in\{0,\frac{1}{n},\frac{2}{n},\dots,\frac{n-1}{n},1\}$, $t=1,\dots,N(m,n)$, selects admissible weight-vectors, i.e., weight-vectors which meet conditions implied by NNN-information $I=II\cup OI$, and gives output information: (1)

average general estimations $\overline{Q}_+(q^{(1)};I),\ldots,\overline{Q}_+(q^{(k)};I)$; (2) standard deviations $S(q^{(1)};I),\ldots,S(q^{(k)};I)$; (3) measures $P(j,l;I)$, $j,l=1,\ldots,k$, $j ? l$, of reliability of revealed preference relation between objects $x^{(j)},x^{(l)} \lfloor X$. These output results are pictorially represented in the form of so called *ASPID-diagram*. Above mentioned quantitative image of qualitative information (numeric image of NNN-information) for weight-coefficients is represented in the form of ASPID-diagram too.

ASPID-3W is well structured and simple in control. But a more experienced user may find in the DSSS an advanced and powerful tool for any interactive multicriteria decision-making with leaning upon uncertain information.

6. DISCUSSION OF THE GENERAL INDICES OF SUSTAINABILITY, WHICH ARE OBTAINED BY APSID – 3W PROGRAM

The analysis is based on the results obtained by the DSSS ASPID–3W exploitation. In this analysis we will take into consideration different alternative as regard weighting factors. In our analysis of the effect of individual criteria on the General Index of Sustainability it is assumed four different cases which will reflect changes in the mutual relation of the weighting factors on the decision making process.

The following cases are taken into a consideration:

1. $I = I_1 = \rightarrow$ - there is no information about admissible weight-coefficients at our disposal (i.e., the set I of corresponding equalities and inequalities is empty set: $I = \rightarrow$).
2. $I = I_2 = \{w_1 > w_2 > w_3 > w_4\}$ - specific criteria (specific indices) are strictly ranked by their influence on General Sustainability Index.
3. $I = I_3 = \{w_2 > w_4 > w_1 > w_3\}$ - we have the ordinal information I_3 about weight-coefficients.
4. $I = I_4 = \{w_1 = w_4 > w_2 = w_3\}$ - we have the ordinal information I_4 about weight-coefficients.

For all four cases input data for DSSS ASPID–3W are the same:

Number of objects (options of the Energy System) k = 4.
Number of initial attributes (sustainability indicators) m = 4.

Number of steps n = 20 (So, it is supposed that the measurement of weight-coefficients is accurate to within the step h = 1/n = 1/20 = 0.05).

Number of all possible variants (of all weight-vectors $w = (w_1,..., w_m)$)
N(4;20) = 1771

6.1 Case 1

In the case, number $N(4,20;I_1)$ of all admissible (from the point of view of NNN-information $I_1 = \rightarrow$) weight-vectors $w^{(t)} = (w_1^{(t)},..., w_4^{(t)})$ is equal with number $N(4,20) = 1771$ of all possible weight-vectors. So, amount $Inf(I_1)$ information I_1 in "bits" ("bit" is a binary unit of information) is equal to zero: $Inf(I_1) = 0$.

The supposition that there is no information about admissible weight-coefficients in our disposal $(I = I_1 = \rightarrow)$ implies equal estimations $\overline{w}_i(I_1) = 0.25$, $i = 1,...,m$, for all weight-coefficients, table 14. But it is not the simple case where we plainly suppose that all weights are equal. No, in our case standard deviations (which are measuring "uncertainty" of the weight-estimations) are taken into consideration table 14.

Table 14. $\overline{w}_i(I_1)$ of weight-coefficients and their standard deviations $s_i(I_1)$, $i = 1,...,4$, for Case 1.

Sustainability Indicators	Weight-coefficients estimations $\overline{w}_i(I_1)$	Standard deviations $s_i(I_1)$
RI	0.250	0.212
EI	0.250	0.212
SI	0.250	0.212
EcL	0.250	0.212

The data, which is contained in Table 14, may be visualized by so called ASPID-diagram, presented on Figure 2. Here the estimations $\overline{w}_i(I_1)$ of weight-coefficients are represented by the marked middles of the "thick" (red) segments, the length of every segment being equal to the doubled standard deviation $s_i(I_1)$. The right end of a "thin" (blue) segment between two "thick" segments, corresponding to r -th and s -th weights, shows the probability $p(r,s;I_1)$.

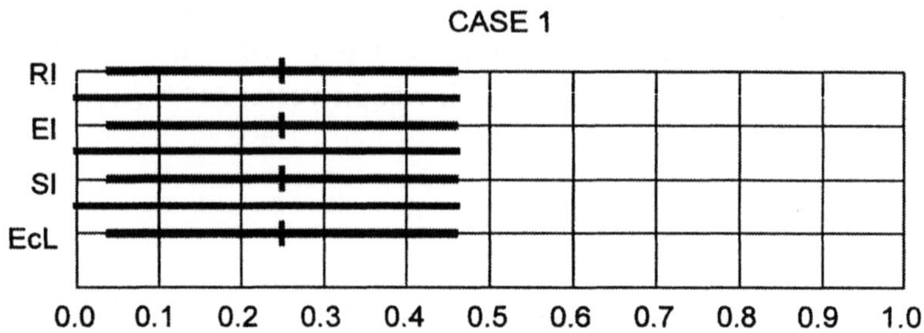

Figure 2. $\overline{w}_i(I_1)$ of weight-coefficients, their standard deviations $s_i(I_1)$, and probabilities $p(r,s;I_1)$ of pair-wise dominance of corresponding random weights $\tilde{w}_r(I_1)$, $\tilde{w}_s(I_1)$ for Case 1.

For Case 1, when there is no information about weight-coefficients at our disposal, formula for general estimation $\overline{Q}_+(q;I)$ transforms into the expression of arithmetical mean

$$\overline{Q}_+(q;I_1) = M\tilde{Q}_+(q;I_1) = M\left(\sum_{i=1}^{m} q_i \tilde{w}_i(I_1)\right) = \sum_{i=1}^{m} q_i \left(M\tilde{w}_i(I_1)\right) = \frac{1}{m}\sum_{i=1}^{m} q_i \tag{16}$$

Values $\overline{Q}_+(q^{(j)};I_1)$, $j=1,...,4$, of the General Index of Sustainability (16) and corresponding standard deviations are represented in the Table 15.

Table 15. General Indices of Sustainability for Case 1.

Options	General Sustainability Index $\overline{Q}_+(q^{(j)};I_1)$	Standard Deviation $S(q^{(j)};I_1)$
Option 1	0.643	0.188
Option 2	0.854	0.114
Option 3	0.333	0.200
Option 4	0.419	0.146

The probabilities $P(j,l;I_1)$ of pair-wise dominance of randomized general indices may be seen in the Table 16. These probabilities may be treated as measures of reliability of the corresponding pair-wise dominance.

Table 16. Probabilities $P(j,l;I_1)$ of pair-wise dominance of General Indices of Sustainability for Case 1.

j \ l	Option 1	Option 2	Option 3	Option 4
Option 1	0.000	0.216	0.823	0.775
Option 2	0.784	0.000	0.999	0.998
Option 3	0.177	0.001	0.000	0.335
Option 4	0.225	0.002	0.665	0.000

The data, which are contained in tables 15 and 16 may be visualized by ASPID-diagram, presented on Figure.3. Here the estimations $\overline{Q}_+(q^{(j)};I_1)$ of General Sustainability Index are represented by the marked middles of the "thick" (red) segments, the length of every segment being equal to the doubled standard deviation $S_j(I_1) = S(q^{(j)};I_1)$. The right end of a "thin" (blue) segment between two "thick" segments, corresponding to j-th and l-th options, shows the probability $P(j,l;I_1)$.

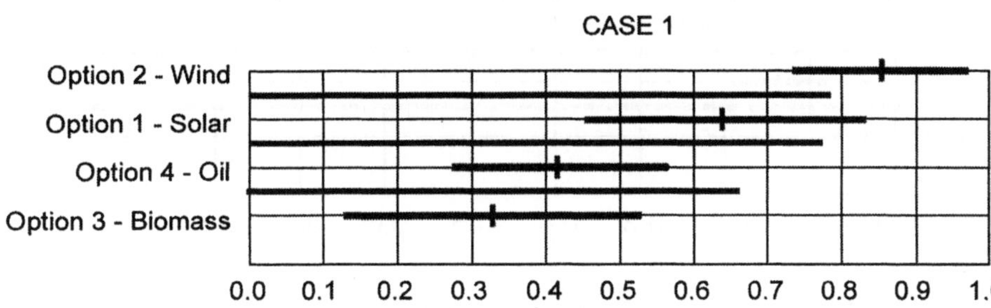

Figure 3. General Indices of Sustainability for Case 1.

From the presented data as shown in Figure 3 it can be concluded that in the Case 1 with $I = I_1 = \rightarrow$, Option 2 – Wind Energy is having the highest value of General Index for Sustainability Assessment. It can be noticed that this option is having minimal value of standard deviation, which represents the accuracy of the obtained rating among the options under consideration.

6.2 Case 2

Case 2 represents an alternative with ordinal information

$$I = I_2 = \{w_1 > w_2 > w_3 > w_4\}, \tag{17}$$

reflecting relation among the individual weighting factors. Now the set $W(4,20;I_2)$ has only $N(4,20;I_2)=47$ elements and we can calculate the amount $Inf(I_2)=5.24\,Bit$ of information I_2.

Table 17. $\overline{w}_i(I_2)$ of weight-coefficients and their standard deviations $s_i(I_2)$, $i=1,...,4$, for Case 2.

Sustainability Indicators	Weight-coefficients Estimations $\overline{w}_i(I_2)$	Standard Deviation $s_i(I_2)$
RI	0.540	0.119
EI	0.284	0.075
SI	0.139	0.060
EcL	0.036	0.043

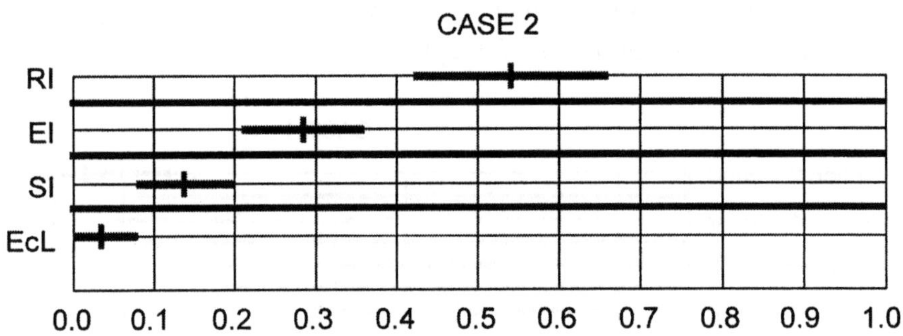

Figure 4. $\overline{w}_i(I_2)$ of weight-coefficients, their standard deviations $s_i(I_2)$, and probabilities $p(r,s;I_2)$ of the pair-wise dominance of corresponding random weights $\widetilde{w}_r(I_2)$, $\widetilde{w}_s(I_2)$ for Case 2.

Value $\overline{Q}_+(q^{(j)};I_2)$, j = 1…4 of general index of sustainability and corresponding standard deviations are presented in Table 18.

Table 18. General Indices of Sustainability for Case 2.

Options	General Sustainability Index $\overline{Q}_+(q^{(j)};I_2)$	Standard deviation $S(q^{(j)};I_2)$
Option 1	0.671	0.040
Option 2	0.962	0.022
Option 3	0.587	0.105
Option 4	0.236	0.070

The $P(j,l;I_2)$ of pair-wise dominance of randomized general indices are shown in Table 19.

Table 19. Probabilities $P(j,l;I_2)$ of pair-wise dominance of General Indices of Sustainability for Case 2.

j \ l	Option 1	Option 2	Option 3	Option 4
Option 1	0.000	0.000	0.830	1.000
Option 2	1.000	0.000	1.000	1.000
Option 3	0.170	0.000	0.000	1.000
Option 4	0.000	0.000	0.000	0.000

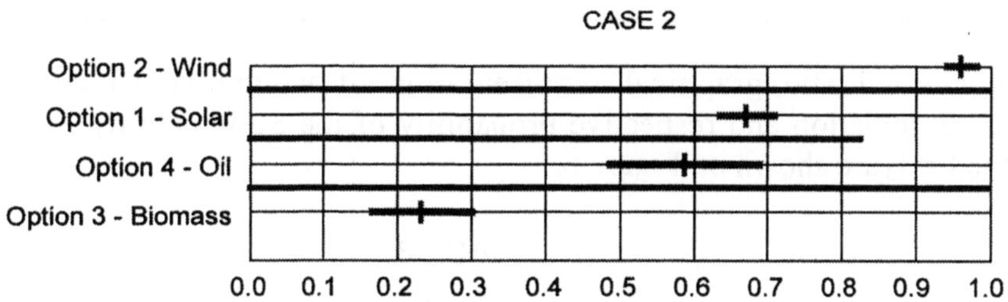

Figure 5. General Indices of Sustainability for Case 2.

Adding an ordinal information I_2 to the procedure for the rating by Sustainability Index the option under consideration does not change priority list but it has added certainty in the selected option by decreasing its standard deviation. Besides this conclusion the ordinal information giving possibility to investigate the importance of the individual criteria in the decision making procedure. For the sustainability assessment of energy systems under consideration it becomes obvious that the selection of Option 2 – Wind Energy is a natural choice which can sustain different criteria shpwn in Figures 4 and 5.

6.3 Case 3

Case 3 represents the alternative for the Sustainability Assessment based on ordinal information

$$I = I_3 = \{w_2 > w_4 > w_1 > w_3\}, \tag{18}$$

giving priority to environment and economic criteria.

This implies that the rating of the criteria is inverse in comparison with Case 2 and priority is given to the environment criteria followed by economy criteria, resource criteria and social criteria.

Table 20. $\overline{w_i}(I_3)$ of weight-coefficients and their standard deviations $s_i(I_3)$, $i = 1,...,4$, for Case 3.

Sustainability Indicators	Weight Coefficients Estimations $\overline{w_i}(I_3)$	Standard Deviation $s_i(I_3)$
RI	0.139	0.060
EI	0.540	0.119
SI	0.036	0.043
EcL	0.284	0.075

Numerical and graphical representation of weight coefficients with standard deviation and respective probability of the Sustainability Indicators for the I = I_3 are shown on Figure 6.

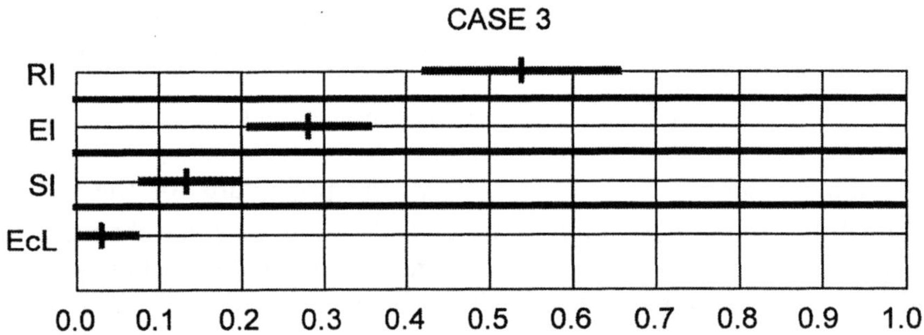

Figure 6. $\overline{w_i}(I_3)$ of weight-coefficients, their standard deviations $s_i(I_3)$, and probabilities $p(r,s;I_3)$ of pair-wise dominance of corresponding random weights $\tilde{w}_r(I_3)$, $\tilde{w}_s(I_3)$ for Case 3.

Table 21. General Indices of Sustainability for Case 3.

Options	General Sustainability Index $\overline{Q}_+(q^{(j)};I_3)$	Standard Deviation $S(q^{(j)};I_3)$
Option 1	0.840	0.042
Option 2	0.839	0.042
Option 3	0.151	0.067
Option 4	0.362	0.023

Table 22. Probabilities $P(j,l;I_3)$ of pair-wise dominance of General Indices of Sustainability for Case 3.

	Option 1	Option 2	Option 3	Option 4
Option 1	0.000	0.489	1.000	1.000
Option 2	0.511	0.000	1.000	1.000
Option 3	0.000	0.000	0.000	0.000
Option 4	0.000	0.000	1.000	0.000

Numerical and graphical presentation of the General Sustainability Index and Probability $P(j,l;I_3)$ for Case 3 is shown on Figure 7 and Tables 21 and 22, respectively.

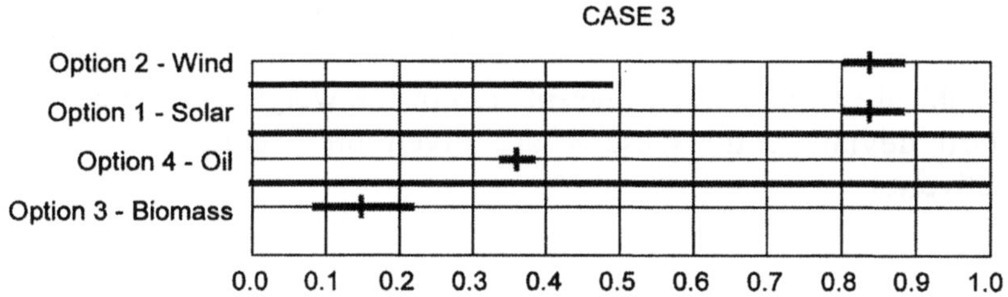

Figure 7. General Indices of Sustainability for Case 3.

With the emphasis on the environment and economic criteria the difference between the option 2 and option 1 is becoming very small so that there is no difference in rating of these two options. It could be noticed that the standard deviation has become smaller and probability for Option 2 has gain substantial confidence.

6.4 Case 4

Case 4 is designed as the alternative to reflect priority to be given to the resource and economic criteria in comparison to the environment and social criteria. These conditions may be represented in the form of the system

$$I = I_4 = \{w_1 = w_4 > w_2 = w_3\} \quad (15)$$

of equalities and inequalities fore the weight-coefficients.

Now the set $W(4,20;I_4)$ has only $N(4,20;I_4) = 5$ elements and we can calculate the amount $Inf(I_4) = 8.47\,Bit$ of information I_2 .

Selection of the weight coefficients relations in this case is determined by the intention to investigate what effect will have priority given the resource and economic criteria in relation to the environment and social criteria.

Table 23. $\overline{w}_i(I_4)$ of weight-coefficients and their standard deviations $s_i(I_4)$, $i = 1,...,4$, for Case 4.

Sustainability Indicators	Weight Coefficients Estimations $\overline{w}_i(I_4)$	Standard Deviation $\overline{w}_i(I_4)$
RI	0.400	0.071
EI	0.100	0.071
SI	0.100	0.071
EcL	0.400	0.071

Graphical and numerical presentation of the weight coefficients and their standard deviation for Case 4 are given on Figure 8 and Table 23, respectively.

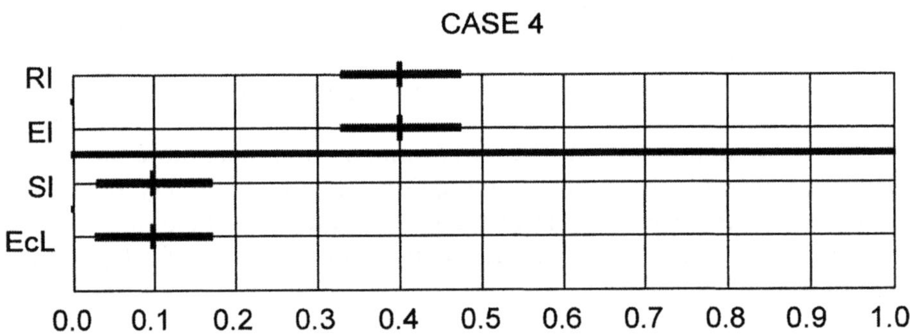

Figure 8. $\overline{w}_i(I_4)$ of weight-coefficients, their standard deviations $s_i(I_4)$, and probabilities $p(r,s;I_4)$ of pair-wise dominance of corresponding random weights $\tilde{w}_r(I_4)$, $\tilde{w}_s(I_4)$ for Case 4.

Table 24. Values of the options' General Indices of Sustainability for Case 4.

Options	General Sustainability Index $\overline{Q}_+(q^{(j)};I_4)$	Standard Deviation $S(q^{(j)};I_4)$
Option 1	0.778	0.063
Option 2	0.767	0.041
Option 3	0.433	0.047
Option 4	0.313	0.050

Table 25. Probabilities $P(j,l;I_4)$ of pair-wise dominance of General Indices of Sustainability for Case 4.

	Option 1	Option 2	Option 3	Option 4
Option 1	0.000	0.600	1.000	1.000
Option 2	0.400	0.000	1.000	1.000
Option 3	0.000	0.000	0.000	0.800
Option 4	0.000	0.000	0.200	0.000

Graphical and numerical presentation of the General Sustainability Index and probability $P(j,l;I_4)$ for the Case 4 are shown on Figure 9 and Table 25 and 26, respectively.

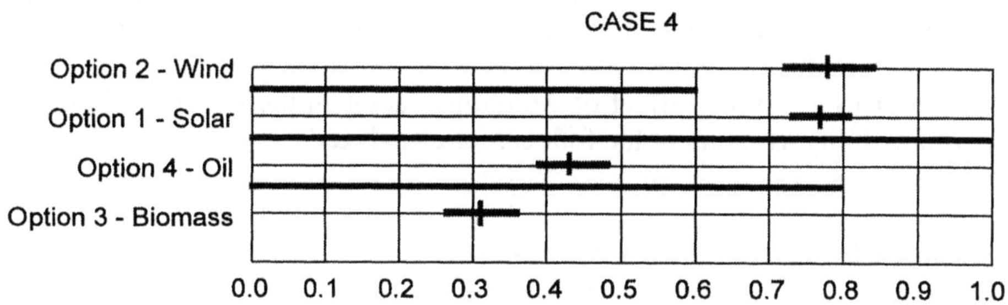

Figure 9. General Indices of Sustainability for Case 4.

Since Case 4 is designed with priority given to the resource and environment criteria the obtained results are suppose to reflect its dominance and significance for the final decision of the priority list. The change in priority is noticed so that option 1 is having higher rating than option 2, but very small advances and reliability. Again, this proves that the selection of option 2 can sustain different priority with sufficient accuracy.

7. DISCUSSION OF THE RESULTS

As it can be noticed from the table 12 the selection of options is aimed to make comparison between the different energy sources used for the production of energy to satisfy the minimum energy need for the undeveloped rural area. The availability of energy sources on the island under consideration are introduced as the input values corresponding to the specific location.

The criteria defined by Fuel resource indicator is aimed to rank options by the amount of organic fuel resources needed to produce a lifetime energy for the respective energy source. From table 12 it can be noticed that the biomass energy option has the maximum value of the Fuel Resource Indicator.

The environment criteria is defined by the amount of CO_2 produced per lifetime energy production. From the table 12 it can be noticed that the wind energy option is having a maximum value if this criteria. This shows that the solar energy option is having the least adverse affect on the environment.

The Job social Indicator is used for the definition of the social criteria for the energy system assessment. Since, the maximum value of social indicator correspond to the wind energy source, it proves that this option is the most favourable option form the social criteria assessment.

The Cost economic indicator is applied for the economic assessment. The oil resource option is having the maximum value of economic indicator and is the first on the priority list of economic indicators.

As it could be noticed the standard deviation for all criteria are very large in comparison with values corresponding to the individual criteria. This will add substantial uncertainty in the assessment of the energy system with the criteria used in this example.

In order to verify the effect of weighting factor in the general criteria for the sustainability assessment the four case are analyzed.

Case 1 – $I = 0$
Case 2 – $I = I_1 = \{ w_1 > w_2 > w_3 > w_4 \}$
Case 3 – $I = I_2 = \{ w_1 > w_2 = w_3 > w_4 \}$
Case 4 – $I = I_3 = \{ w_1 = w_2 > w_3 > w_4 \}$

For the Case 1, from the table 15, it is obvious that the wind energy option has a maximum value of General Sustainability Index with assumption of same weighting factor for all criteria under consideration. For the Case 2 even more the selection of wind energy option is emphasized as it can be seen from the table 18. It shows only that the standard deviation of General Sustainability Index is less pronounced. In the Case 3, although weighting coefficient is giving substantial priority to the environment criteria the General Sustainability Index for solar and wind energy are the same as show on table 21. The Case 4 shows that under certain constrain the priority list can be changed table 24.. Due to the high value of resource weighting coefficient the solar option has obtained priority

With non-numeric information related to weighting factors mutual relation, it will be possible to obtain the better quality of the assessment procedure. Four demonstrated cases are arbitrary selected to demonstrate the effect of the weighting factors relation to the decision of the priority among the option under consideration. It should be mentioned that the analysis of the effect of non-numerical information on decision making process in selecting specific option for the energy system may lead to the remarkable confidence in the decision making process.

Sustainable energy development is an ultimate goal of modern society in order to meet ever growing demand for the new energy resources. In particular it was recognized that the complexity of the global system will require special attention to the interaction between life support systems.

It was demonstrated that there is possibility to define the consistent set of sustainability indicators to be used in the assessment of energy system. In this respect, it was presented four group of the indicators which reflect the resource, environment, social and economic criteria.

Due to the deficiency of reliable data for the sustainability assessment, the demonstration of the methodology for sustainability assessment of energy system is based on the simple energy system with limited number of indicators. The selection of energy resources for the small island has shown the application of decision making procedure based on the non-numeric information about the weighting factor for the individual criteria.

8. REFERENCES

1. Afgan, N.H., Al Gobaisi, D., Carvalho, M.G., Cumo, M., Sustainable Energy Management, Renewable and Sustainable Energy Review, 2, pp. 235-286, 1998.
2. Agenda 21, Chapter 35, Science for Sustainable Development, United Nations Conference on Environment and Development, published by the Centre for Our Common Future, Geneva, Switzerland, 1992.
3. Geis, D., Kutzmark, T., Developing Sustainable Communities: The Future is Now, Public Management Magazin, International City/County Management Association, Washington DC, 1998.
4. Voinov, A.A., Paradox of Sustainability, Institute for Ecological Economics, Solomons, MO 20686, USA, 1995.
5. Pearce, A., Turner, R.K., Economics of Natural resources and the Environment, Harvester Wheatsheeaf, London, 1990.
6. Becker, E., Jahn, T., Stiess, I., Wehling, P., Sustainability: A Cross-Disciplinary Concept for Social Transformations, MOSCT Policy papers 6, UNESCO, September 1997.
7. Gottchalk, Ch.M., Industrial Energy Conversion, John Wiley & Sons Ltd. West Sussex, England, 1996.
8. Markvart, T.T., Solar Electricity, John Wiley & Sons Ltd.West Sussex, England, 1994.
9. Walker, J.F., Jenkins, N., Wind Energy Technology, John Wiley & Sons Ltd. West Sussex, England, 1997.
10. Larson, E.D., Williams, S.F., Biomass gasifier steam- injected gas turbine co-generator, Journal of Engineering for Gas Turbine and Power, 112, pp. 157-163, 1990.
11. Werko-Brobby, Ch.Y., Hagen, E.B., Biomass Conversion and Technology, John Wiley & Sons Ltd. West Sussex, England, Dahl K., 1996.
12. Hovanov, K.N., Hovanov, N.V., DSSS "ASPID–3W". Decision Support System's Shell "Analysis and Synthesis of Parameters under Information Deficiency - 3-d modification for Windows". "ASPID-3W" registered by Federal Agency for Computer Programs Copyright Protection (Russia Federation, Num. 960087), 22 September 1996.
13. Hovanov, N., Kornikov, V., Seregin, I., Qualitative information processing in DSSS ASPID-3W for complex objects estimation under uncertainty, Proceed. Int. Con. "Informatics and Control". St. Petersburg (Russia), P.808-816, 1997
14. Hovanov, N., Analysis and Synthesis of Parameters under Information Deficiency. St. Petersburg. St. Petersburg State University Press, (Monograph, in Russian), 1996.
15. Afgan, N., Carvalho, M.G., Hovanov, N., Energy System Assessment with Sustainability Indicators, Energy Policy, (accepted for publication)

CHAPTER 6

SUSTAINABILITY ASSESSMENT OF HEAT EXCHANGER DESIGN

1. INTRODUCTION

Assessment of the thermal energy system has been under consideration from the very beginning of its use. In particular, heat exchangers have been always optimized to obtain maximum heat rate per unit surface. Most of approaches developed have been based on the thermodynamic assessment of the respective design [1]. A thermodynamic optimization is aimed to minimize the thermodynamic inefficiency and is a part of the design optimization. Energy-based techniques were commonly used in the optimization procedure for the design of heat exchangers. The pinch analysis is usually applied to a heat exchanger network by identifying the process pinch and guiding the heat exchanger network design and its integration with other components[2]. The pinch method contains elements of a simplified exergy analysis [3]. It is no doubt that the future design of heat exchangers will strongly relay on these methods and will represent an essential tool for the parametric study of any heat exchanger design.

Since the sustainability concept requires wholeness in the approach, it is of interest how this approach will reflect in the design of heat exchangers. In this respect it is of paramount importance to be able to define the design criteria.

There has been a number of attempts to define the criteria for the assessment of sustainability in commercial products. In this respect, the Working Group of UNEP on Sustainable Development [4] has come out with qualitative criteria for the assessment of product design.

We would like to introduce these criteria for specific application to heat exchanger design. In this context, heat exchanger design is taken as the entity, which should comply with the sustainability criteria.

Heat exchanger design is considered to encompass the following aspects:

1.1 Strategic Design

Strategic design will require holistic planning, that meets and considers all interrelated effects e.g. logistic, space planning, and resource planning.

Regarding the energy system, it may be interpreted as: mixed energy concept with optimization of local resources, urban and industrial planning with transport optimization, and the use of renewable energy sources.

1.2 Optimized Design

The design optimization of the heat exchanger, means the selection of structure and design parameters of a system, so as to minimize energy cost under constraints associated with available material and financial resources, environment and government regulations, as well as attention to safety, reliability, availability, and maintainability of the system.

1.3 Dematerialization of Design

This will imply that the energy system, plant and equipment, are designed for optimal use of information technology, in order to prevent duplication and operational malfunction, as well as to assure rational maintenance scheduling. Dematerialization in the design, may be seen as the introduction of knowledge based systems, the use of virtual library, digitized video, on-line diagnostic systems, and the development of new sensor and new combustion technologies.

1.4 Longevity of Design

Complex energy systems are commonly composed of different subsystems and individual equipment elements including heat exchangers. It has been recognized, that the life time of elements and subsystems are not equal. In this respect, the optimal selection of life cycle for elements and subsystems, may lead to a retrofitting procedure, which will reflect the need for the application of sustainable criteria. Examples for this criteria can be seen in the modular design of the subsystems, standardization of the elements, lifetime monitoring, and assessment and coordination of suppliers and buyers.

1.5 Life Cycle Design

This will mean that the energy system and its subsystems including heat exchangers have to be designed to meet sustainability through every stage of the life cycle. It is known that the energy system is designed to work under different conditions, in order to meet load change, environment change, social change, etc. It is obvious that there will be different cycles, for each of the mentioned time scale processes. In this respect, the system has to fulfill its function, without failing to meet sustainability requirements.

2. SUSTAINABILITY INDICATORS FOR HEAT EXCHABGER DESIGN

It is known that any numerical number, semantic expression, or mathematical sign, is information. Examples of this type of information, can be found in any reading of temperature, or other time dependent variables, describing the state of a system. Information is also a parameter used in the design of the respective system. Also, positive or negative sign of the variable is information [5].

Collecting and processing information will convert it into data. So, data represents agglomerated information, which is partially or completely processed. Examples of data can be found in a parameter, which describes evaluated information, to be used for a specific purpose. In this respect the average inlet temperature of cooling water in the condenser is data obtained by an averaging procedure, adapted to this purpose. The heat transfer coefficient, used in the design of a condenser for heat transfer evaluation, is also data obtained by an experimental procedure.

In order to use data for the assessment of the respective system, it is necessary to convert it into an indicator. The indicator represents the measuring parameter for the comparison between the different states or structures of the system. As an example, the efficiency of the system is the indicator for the quality of energy use, in that respective system. We can also evaluate the different structures of the systems, by the indicator representing respective entities of the system. In this context, the respective indicators measure the assessment of the use of intelligence in the improvement of the system compatibility with its surroundings.

Element indicators are defined in order to quantify criteria [6], for the sustainability assessment of any design of a thermal system. In this respect, technology development and the efficient use of resources, is of fundamental importance. It is obvious that the efficient use of energy resources is a short-term approach, which may provide beneficial returns in the near future. For technology development however, long term research is required. In some cases, social adjustment is required in order to meet requirements of the new energy sources.

Potential improvements in the energy conversion process is a driving force for its development. The conversion process is a promising tool for the analysis of an energy system. The exergy analysis[7] is based on the maximum potential availability and its use for the assessment of the conversion process. By definition, exergy is a parameter for the validation of the efficiency of the energy conversion process and system. Taking into account the laws of thermodynamics, technology improvement appears as a significant factor, responsible for an entropy change in energy systems. The application of Carnot's principle, therefore allows the determination of an absolute e limit to any transformation of a deposit of free energy.

2.1 Sustainability Assessment Parameters

In order to facilitate sustainability assessment, indicators are used as parameters for the measurement of sustainability. In this respect, attention is focused on resource indicators and environment capacity indicators. Each of these entities has to be determined with respect to specific parameters, which can be used to calculate the characteristic indicators, for the assessment of their respective entities. The characteristic indicators are:

a. Possible change indicator: The possible change indicator characterises changes, with respect to the maximum change technically possible. If applied to the energy resource, it is the difference between known resources exploitable with known technology, and the current resource to be obtained with the present technological capacity. As an example, it can be shown that total coal resources in 1980 were 2000 Mtons, and in 1985 were 2300Mtons. The maximum potential resource to be used without affecting the available resources is 300 Mtons. The indicator for possible change is defined as the ratio of the maximum potential resources and the average value for the period under consideration. So that the Possible Change Indicator (PC1) is 0.13. The possible change indicator is a function of the technology development in the discovery of resources. This implies that there are two ways to increase the availability of the resources, namely; additional investment for new discovery and new technology for the resource exploration.

b. Current change indicator: The current change indicator is defined as the resource consumption change of the respective resource. It can be expressed as the difference in current consumption change, in the reference time period. Applying the measurement of this indicator to the period 1980 - 1982, it can be seen that Current Change Indicator (CCI) for coal is 0.467. It is known that the current consumption of the energy resources strongly depends on the efficiency of energy use. The possible change in the efficiency of energy use, includes firstly, the efficiency of the primary energy source conversion, and secondly, the efficiency of the final energy use.

c. Resource indicator: By definition, the Resource Index (RI) is the ratio between the Current Change Indicator and the Possible Change Indicator. In essence, it will represent a qualitative measure of resource depletion. It may also be interpreted as the resource scarcity change due to resource depletion. For the case of coal resources, the Resource Indicator is 0.065.

d. Financial effect of resource use: The resource indicator reflects the material balance of the respective primary energy resource. In order to use it for assessment of the quality of primary energy use, it has to be connected with the financial effect resulting form its use.

2.2 Resource Indicator

The resource indicator can be defined as the amount of material used to perform specific functions. Material is used for the construction of the specific equipment and the material is used as the energy resource needed for a respective function. Each element of thermal equipment represents some metal used for construction that is obtained by processing raw material from available resources to satisfy the form defined by the material function. In the processing of resource material adequate energy is required to be spent, to convert it to its final form. This energy is called production energy. Since every element is designed to perform a function related to a specific operation in the system; energy is needed to perform respective functions within the system.

In order to define an indicator for the use of resources, for an element of the energy system, let us represent them in following form,

$$R_m = \frac{G_m}{E}$$

where

G_m - weight of material

E - energy transformation

In order to define its relative value, which may be used for qualitative assessment of a specific element, we can write,

$$\rho_m = \frac{R_m}{R_{m,opt}}$$

where,

$R_{m,Opt}$ - is optimal resource consumption for specific element

In this case, the subscript m, reflects materials used for production, and include metals, and energy resources.

If it is assumed, that the energy use for production of the respective element is,

$$E_p = (E_m + E_{prod})R_m$$

where,

E_m and E_{prod} are the energy consumptions for material production and material formation, respectively.

With the assumption that the life time of an element is T_e, the energy used for that element to perform its function is _, and the total energy consumption for the element function is,

$$E_t = E_p + \xi * \tau_e$$

The energy resource consumption for the total energy consumption of the respective element is,

$$R_e = \frac{E_e}{r_e}$$

where,

r_e - specific energy produced by the respective energy resource

The relative value of energy resource is,

$$\rho_e = \frac{R_e}{R_{e,opt}}$$

Following this methodology, it can be concluded, that for total resource consumption, the indicator is composed of the individual indicators reflecting their respective material usage in the design of an element and energy resource indicator.

2.3 Environment Capacity Indicator

The second criterion for sustainability assessment is based on the environmental capacity consumption with the use of respective element design.

The optimal energy consumption is obtained using the Carnot cycle, with respect to the environmental capacity consumption. If we take the ratio of the obtained environmental capacity, to its optimal value, we will obtain,

$$\varepsilon = \frac{C}{C_{opt}}$$

where

C and C_{opt} - environment capacity and optimal value, respectively

ε - environment capacity efficiency

The optimal environmental capacity value can be defined for every design of an energy system element, so as to include all potential pollutants. Since, the environmental capacity efficiency is a function of the number of parameters describing a specific design, it is of special interest to determine the value, which corresponds to the design. This will represent the environmental capacity indicator, which will be used for the assessment of a specific design.

3. DEMONSTRATION OF SUSTAINABILITY CRITERION FOR HEAT EXCHANGER DESIGN

For many years in the power industry, condensers were considered to be the heat transfer elements, which best represent the heat exchanger family [8]. There are three numerical procedures used for the calculation of heat transfer and fluid flow in a heat exchanger. The first stage in the design of a condenser is the determination of the heat rating and the heat transfer surface area, which includes sizing of the tube diameter and the tube length. The second stage involves the evaluation of a two-dimensional study of two-phase flow in the shell side of the condenser. The third stage is comprised of a detailed three dimensional calculation, of the velocity and temperature fields, in the shell and tube sides of the heat exchanger.

For this exercise, the first stage of the procedure is used, with the aim of defining the respective heat transfer surface area and pressure drop specifications, required to attain stable functioning of the heat exchanger [9].

For demonstration purposes, the power plant condenser considered, has a total heat rating, Q, of 700 MW, a vapour flow, G, of 959 tons/h, with pressure, P of 3.34 kPa, and a respective saturation temperature, T, equal to 26.8_ C. The water inlet temperature, t equals 16 _C and outlet temperature, t is 24.8 _C., and the water mass flow rate G is 55 934 t/h.

Table 1 shows the options taken into consideration, namely; a clean condenser without a fouling factor, half fouling resistance case, and a total fouling resistance value as proposed by TEMA procedure. Table 1 also indicates that the change in the tube diameter and tube material is evaluated.

Table 1. Change in the tube diameter and tube material.

R	d [m]	Material
R = 0	0.028	90 Cu +10 Cu-Ni
	0.026	
R= TEMA	0.024	Carbon Steel

3.1 Sustainability Assessment of Condenser Design

As indicated in Table 1, the cases considered in the analysis of sustainability assessment, consider the effect of several parameters, including the fouling margin effect, the change in tube diameter, and the use of different materials for condenser tubing. The aim of this assessment is to verify changes in the sustainability indicators, and show their effect on the global contribution to resource usage and future development. Figure 1 shows schematic presentation of condenser design.

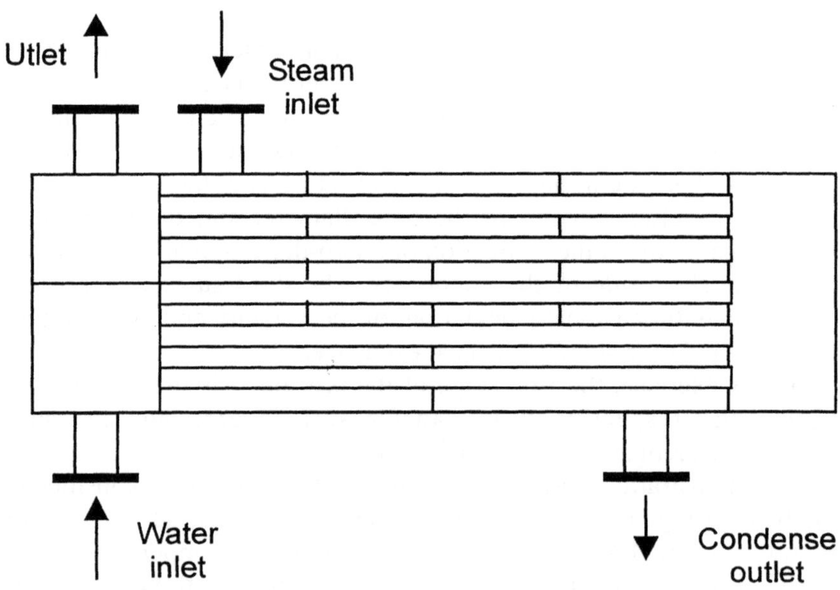

Figure 1. Schematic presentation of condenser design.

The fouling phenomenon is inherent to heat exchanger operation [10]. In particular, fouling in condensers, requires special attention at the design stage. It is commonly taken into account in the design procedure, with respect to the heat transfer rating and heat transfer surface determination. The fouling margin is expressed in form of a fouling resistance in heat transfer calculations, resulting in over-sizing of heat transfer surfaces. This approach to condenser design results in excessive use of natural resources and, at the same time, also contributes to excessive usage of the environment capacity.

The respective indicators measure the fouling effect on the sustainability criteria, and for this purpose, the following indicators are used;

3.2 Construction Material Indicators

By definition the construction material indicator is the ratio of the difference in the weight of the material required for a specific design of a condenser with and without fouling margin and the energy used in the transformation of the material. The heat transfer surface sizing method is used to determine the difference in condenser weight. Attention has been focused only on the tube material [11]. As shown in Table 1, three tubes diameters are taken into consideration, copper alloy and carbon steel are used for the tubing material. For each of these combinations, the clean and fouled surface sizes are calculated.

Results have been presented in the form of the possible change, current change, and resource change indicators for the construction material.

Table 2 shows the data obtained for the cooper alloy and carbon steel tube materials.

Table 2. Cooper alloy and carbon steel tube materials.

D	G [t]		Δ G [t]	Δ G/E [t/Mwe]	PCI	CCI	RI
	R = 0	R = TEMA					
0.028	189,5	256	66.5	0.221	0.350	0.5	0.175
0.026	188.5	248	59.5	0.198	0.315	0.5	0.157
0.024	185	262	77	0.256	0.416	0.5	0.208

D	G [t]		Δ G (t)	Δ G/E [t/MWe]	PCI	CCI	RI
	R = 0	R = TEMA					
0.028	183	251	68	0.226	0.370	0.5	0.180
0.026	180	226	46	0.153	0.250	0.5	0.125
0.024	177	231	64	0.213	0.359	0.5	0.180

3.3 Energy Resource Indicator

This indicator reflects the effect of fouling, on the energy resource consumption, for a specific design of a condenser. The two major contributors to this indicator are the energy (coal) used in converting the natural resource to the tube material and the energy required to pump the coolant during condenser operation, with an assumption of a 30 year lifetime. The energy resource indicator is expressed in form of the possible change, the current change, and the energy resource indicators, if the fuel under consideration is coal.

Table 3 shows the data obtained for copper and carbon steel tubing resource indicators.

Table 3. Copper and carbon steel tubing resource indicators.

Material	$\Delta G_{coal}[10^6\ t]$	(PCI)	CCI	RI
Cooper Alloy	15	$0.57x\ 10^{-3}$	0.5	$0.285x10^{-3}$
Carbon Steel	12	$0.43x\ 10^{-3}$	0.5	$0.215x\ 10^{-3}$

3.4 Environment Capacity Indicator

In this analysis only the CO_2 contribution, due to fouling, is used for the assessment of the environmental capacity indicator. As in the energy resource indicator, the contribution to the environmental capacity indicator is divided in two parts. The first part represents the contribution due to the material production process, and the second part results from the life-cycle energy requirement for system operation. The environmental capacity indicator is expressed in form of the possible change, current change, and resource indicators for CO_2.

Table 4 shows the data obtained for the environmental indicators for copper and carbon steel condenser tube materials.

Table 4. Environmental indicators for copper and carbon steel condenser tube materials.

Material	$\Delta G_{CO2}[10^6 t]$	$(PCI)_{En}$	$(CCI)_{En}$	$(RI)_{En}$
Carbon Steel	6.0	$0.57\ x\ 10^{-3}$	0.5	$0.265\ x\ 10^{-3}$
Cooper Alloy	4.8	$0.43\ x\ 10^{-3}$	0.5	$0.215\ x\ 10^{-3}$

4. DISCUSSION OF RESULTS

The exercise presented in this chapter is meant only to indicate the possible use of sustainability indicators as a tool for the assessment of condenser design. Fouling is considered to adversely affect the condenser operation. Since the condenser is a power plant element, which substantially affects the total power plant efficiency, it is of considerable interest to demonstrate the application of these sustainability assessment factors. Also, the effect of fouling in the condenser on the sustainability indicators, is not time dependent, and thus allows the possibility of using the available data, obtained using the standard design procedure. It should be emphasized that these analysis do not include the change of global resource indicators with time.

It could be noticed that the tube diameter change effect on the sustainability indicators is more pronounced then the change in the tube material. In particular this will be dominant if the total global resource indicator will be used in this evaluation. Since, the global steel resources are for several order of magnitudes larger than the cooper alloy resources. It is obvious that the change to carbon steel condenser design may prove to be less sensitive to the fouling effect on the sustainability indicators.

The energy resource indicator for the carbon steel shows the advantage of using this material, and how that would contribute to a lower usage of energy resources. In particular, it may be shown, that with the use of oil as an energy resource, this effect may be even more pronounced. If, in this analysis, the scarcity forecast of oil resources is used; the use of carbon steel may prove to be beneficial for the design of future power plants.

The environmental capacity is usually measured by the CO_2 contribution to the global atmosphere. In this respect, the assessment of the fouling does not best represent the use of environment capacity indicator. It may be shown, that in the total CO_2 production, the part resulting from material production is preceding the design of condenser, and the part resulting from operation will be spread over the time period during which the fouling process takes place. This may lead us to the conclusion that carbon steel is more favorable as a material for the power plant condenser tubing, with respect to sustainability.

5. REFERENCES

1. Kakac, S., Liu, H., Heat Exchanger Selection, Rating and Thermal Design, CRC Press, Boca Rota, USA, 1997.
2. Linnhoff, B., Pinch Technology for Synthesis of Optimal Heat and Power Systems, Journal of Energy Resources Technologies, 111, pp. 137-144, 1989.
3. Szargut, J., Moris, D.R., Steward, F.R., Exergy Analysis of Thermal, Chemical and Metallurgical Processes, Hemisphere Publishing Corp., Washington, 1976.
4. UNEP Working group on Sustainable Development, Internet Communication, 1997.
5. Bowerman, D.P., Walker, J., Collins, M.C., Indicators of Sustainability, Thomas Jefferson Planning District Commission, 1997.
6. Indicators of Sustainable Engineering, Physical Sciences Research Concil, 1996.

7. Kotas, T.J., The Exergy Method of Thermal Plant Analysis, Krieger Publishing Company, 1995.
8. Bell, K., Heat Exchangers with Phase Change, Heat Exchangers Sourcebook, Ed.J.Palen, Hemisphere Pub. Corp, Washington, 1986.
9. Heat Exchanger Handbook. Ed: G.F. Hewitt, Begell House, Inc., Publishers, New York, USA, 1998.
10. Taborek, J., Assessment of Fouling Research on the Design of Heat-Exchange, Fouling Mitigation in Industrial Heat-Exchange Equipment, Ed. Punchal, Begell House Inc, 1997.
11. TEMA, Standerd of Tubular Heat Exchanger Manifacturers Association, New York, 1978.

CHAPTER 7

WATER AND ENERGY CO-GENERATION ASSESSMENT

1. INTRODUCTION

Water resources have emerged as a key element of sustainable development [1]. Among current users of water, agriculture accounts for the largest share, followed by industry and households. The growing future needs will exacerbate the already serious shortfalls in investment, as well as other weakness such as poor sector efficiency and inadequate prices. The three main elements of sustainable development are economic, environmental and social. Successful and sustainable development can be achieved only by a rational and systematic integration of water-related economic, environmental, social and technical considerations within the broader frame work of a sustainable development strategy. In this respect the recent emergence of transnational and global environmental concerns suggest that water resource analysis should develop an even wider perspective. There are several technology options for water production of which desalination is the most promising.

It is of great interest to treat water desalination processes as an holistic problem including the sustainability assessment based on the criteria implying the resource, environmental, social and economic indicators. From the technical point of view there are a number of desalination processes presently in use on the industrial scale [2]. It has been shown that about 60% of the worldwide desalting plants capacity are located in the Arabic Peninsula. In principle there are three main processes, namely: multi-effect destilation, multi-stage flashing and reverse osmoses. There have been a number of attempts to evaluate advantages of each of the processes taking into consideration different criteria [3,4]. Most of these studies are devoted to the economic aspect of desalination plants. Recently, some attention was given to the environmental aspect of those plants. But still there is a need for the complex approach which will reflect the sustainability aspects of the desalination processes taking into consideration resource, environmental, social and economic aspects with generalised criteria for the priority assessment.

This chapter is devoted to the desalination processes analysis taking into consideration sustainability indicators defined to meet the criteria reflecting different criteria for the sustainability assessment of the respective plants. In particular attention is focused on the assessment of the different energy sources to be used for the desalination processes. In this respect, it was taken into consideration the following options: single purpose MSF plant, dual purpose MSF plant, reverse osmoses plant with electric power form local network and reverse osmoses plant with solar energy.

2. DESALTING PLANTS IN THE GULF AREA: A NEW OUTLOOK

The countries of the Gulf area depend on desalting seawater to satisfy their needs for potable water. In these arid areas, where temperature easily reaches 50°C most of the summer days in Kuwait for example, more than 70% of their power productions are used to operate air-conditioning systems. The continuous rising of standard of living increases the demands for more power and desalted water. The electric power generation and desalted water production in large quantities can be realised by using one of the following options:

1. Separate power plants, and separate desalting plants.
2. Dual Purpose Power-Desalting plants using steam turbines and MSF desalting systems.
3. Separate Power plants and Reverse Osmosis Desalting (using only mechanical energy) systems.
4. Reverse Osmosis with PV solar energy system.

These four options are compared in this study in reference to sustainability indicators with practical examples from Kuwait. The evaluation will be focused on the commercially available desalination technologies.

Most desalinating plants operating in the Gulf countries, consist of steam turbines, each having 300 MW electric power capacity and combined with two units multi stage flash desalination (MSF) units of 6 MGD, (million imperial gallons), each. Steam of moderately low pressure and temperature, (compared to the turbine throttling conditions), is extracted from the turbine to the two MSF units through a cross over pipe between the low and medium pressure turbines to provide the thermal energy needed. Each one million imperial gallon per day (MGD) is consuming about 15 MW thermal energy. Moreover, about 5 kWh mechanical work per m^3 is needed for pumping.

2.1 The MSF Desalting System Energy Consumption

The multi stage flash desalting system, see figure 1, consists of heat input section (HIS), heat recovery section (HRS), and heat rejection section (HJS). The system is a big consumer of both thermal and mechanical energy. Thermal energy may be supplied directly from a boiler to the system. This practice is limited to only small units. This method utilizes high available fuel in producing low availability steam (at moderately low pressure and temperature) suitable for supplying to the MSF desalting system. Large MSF desalting units in the Gulf countries are combined with co-generation steam turbines. Steam supplied to the desalting units is generated at high temperature and pressure, and is expanded in the steam turbine to the conditions required by the desalting system. Thus the steam is producing work before its supply to the desalting units. So, the amount of fuel supply to the steam generator of a co-generation power desalting plant to produce both the power output and desalted water should be allocated between these two products. A rational basis for allocation is adopted here, but first let us see the amount of thermal energy consumed by the MSF desalting system. The multi stage flash desalting system is rated, from energy point, by different methods. The first is the Gain Ratio defined by the amount of desalted water in kg produced per kg of the supplied steam. As an example, all the desalting plants in Kuwait have a design gain ratio of 8. If steam added to the HIS of the desalting unit is saturated vapor at 100°C leaves as condensed saturated liquid at the same temperature, then the energy consumed per unit mass of the desalted water is equal to the latent heat divided by the gain ratio (which is 8 here). So the thermal energy consumed to produce one kg of desalted water is 282 kJ/kg. More thermal energy is consumed to operate a steam ejector used to reject non-condensable gases from the system stages usually operated at pressure below the atmospheric pressure. The thermal energy used by the ejector is close to 16 kJ/kg of desalted water but at much higher pressure than steam supplied to the HIS. So the total thermal energy supplied to produce one kg of desalted water is in the range of 300 kJ/kg. If the desalting system is directly operated by a boiler of 80% efficiency, then the fuel energy consumed to supply its thermal energy need is 375 kJ/kg of desalted water.

Moreover, the operation of the MSF pumps (namely re-circulation, cooling water, product, condensate, and brine rejection pumps) consumes mechanical energy. In case of Kuwait, the mechanical energy consumption in most of Kuwait plants is in the range of 16-20 kJ/kg. When this mechanical energy, is generated from a power plant having 0.35, thermal efficiency, the fuel energy required to produce 18 kJ/kg is 50 kJ/kg. So the total fuel energy needed to produce 1 kg of desalted water from MSF directly driven by boiler is 375 + 50 = 426 kJ/kg.

In practice, large MSF is using bled steam from a steam turbine to supply its HIS with the required thermal energy. A lay out of a Co-generation power desalting plant is given in figures 2a and 2b with and without steam extracted to the desalting units. The shown plant is combined to two MSF desalting units producing 12 MGD (each 1 MGD = 4550 m^3/d). The two desalting units, producing 54600 m^3/day, consume 180 MW thermal energy supplied their HIS (285MJ/m^3), 10 MW thermal supplied to their steam ejectors (15.8MJ/m^3), and 10 MW mechanical energy (15.8MJ/m^3) to operate their pumps. These figures show that when the plant is producing 225 MW power only, the steam generator output is 581.4 MW thermal energy; and when it produces 225 MW power plus 12 million imperial gallons (MGD) of desalted water, the steam generator output is 671.65 MW thermal energy. This means that in order to supply 180 MW thermal energy to the HIS of the two desalting units, the steam generator has to increase its output by only 90.25 MW. This almost 50% saving compared to the directly boiler driven case. So, the minimum fuel energy to be charged to the desalination process is 90.25/0.9 = 100.27 MW to account for the heat input section thermal energy, where 0.9 represent an average efficiency of the boiler. More fuel energy should be added to account for the thermal energy consumed by the steam ejectors (10/0.9 = 11.11 MW), and the pumping mechanical work (10/0.35 = 28.6 MW, where 0.35 is power plant efficiency).

More rational method of evaluating the fuel energy to be charged to the desalting units is to find the equivalent work loss due to the extraction of steam from the turbine to the HIS instead of expanding this steam to the condenser. If it expanded to the condenser, it would produce 40.65 MW mechanical work. This is equivalent to 40.65/0.35 = 116.14 MW fuel energy. By using this last method, the fuel energy charged to produce 12 MGD, and accounts for all heat added to heat input section, and ejectors plus the work consumed by the pumps is 116.14 + 11.11 + 28.6 = 155.85 MW. This equivalent to 246.6 MJ fuel energy per m^3 desalted water or 86.32 MJ mechanical energy per m^3 desalted water (= 23.98 kWh/m^3 mechanical work).

The average rate of desalted water production (and consumption) in Kuwait was 1997 is 200.778 MGD, i.e. 333.442 million m^3 annual production. For a fuel oil of 40000 kJ/kg, the fuel energy consumed to produce this amount of water was 2055.67 million tons per year.

Surprisingly enough, if a Reverse Osmosis desalting plant is used, with maximum guaranteed energy consumption of 7.5 kWh/m^3 product water (27 MJ mechanical work/m^3 product, or 77.14 MW fuel energy/m^3), the annual fuel energy consumption would be 643.07 millions tons per year.

To complete the picture, the desalted water produced by MSF desalting units driven directly by boilers requires a fuel energy consumption of 437.7 MJ/m^3. For the Kuwait case, the daily production of 200.778 MGD requires 3648.7 million tons of fuel.

The MSF Hardware

Recent multi stage flash (MSF) desalting units used in Kuwait have production capacity of 6 MGD each. Design parameters are affecting the required number of stages, heat transfer area, plant dimensions, stream flow rates, pressure drops, pumping energy, ...etc. A design example is presented here for 6 MGD unit operating under the following condition:

1. Top brine temperature is 90 °C,
2. Gain ratio is 8,
3. Maximum and minimum seawater temperatures are 32.2 °C and 14.5 °C respectively,

By assuming a unit of cross flow tube arrangement in the stages, and making heat and mass balances, the following flow rates of streams are resulted. Distillate (water production), D = 1127.7 tons/h at 38.6°C, steam to brine heater, S = 141 tons/h at 100 C, re-circulation brine at the brine heater outlet, R = 14286 tons/h at 90.56 C, brine blow-down, B = 1798 tons/h at 40.5 C, Make-up (feed), F = 2925 tons/h at 40.23 C, and cooling seawater, M = 6269.3 tons/h at 28 C.

Heat Transfer Areas

Parametric analysis was conducted and showed a trade-in between increasing the number of stages and decreasing the heat transfer areas or vice versa (see Figure 3). Twenty four stages were chosen here, and the following heat transfer areas and materials are resulted:

1. Brine Heater, (Heat Input Section, HIS): heat transfer surface area of 3580 m^2 (1378 tubes of 44.45 mm O.D., and 18,606 mm length); Cu-Ni (70:30); 1.22 mm thickness and 2013 m/s brine velocity. The area calculated was based on overall heat transfer coefficient, U = 2.1 $kW/m^2.K$ and 3.39 °C logarithmic mean temperature difference.
2. Heat Recovery Section (HRS):

   (ii) (a) Heat recovery section (HRS): the tubes of the first to third stage are chosen to be Cu-Ni (70:30) and the condenser area of each stage is 3714 m^2 (44.45 mm O.D. × 1451 tubes of 18,330 mm length). (b) The tubes of the 4-21 stages of the HRS is chosen as Cu-Ni (90:10); and the tubes area of each stage is 3714 m^2 (same number and dimensions as before).

(iii) Heat rejection section (22-24 stage): The tubes of these rejection stages are made of Cu-Ni (70:30), with 3230 m^2 surface area per stage (44.45 mm O.D. × 1243 tubes of 18,330 mm length). The material choice was based on the material chosen for modern Makka-Taif MSF plant [1].

Thus a total heat transfer surface area of 92,000 m^2 is needed, including 735 m^2 for the condenser system. The price of 1 m^2 Cu-Ni (90:10) tubing is estimated as US$ 84/m^2 (a 50% price increase compared to that stated by Volta as of 1984 [2]). So, the tubing cost is US$ 7.73 × 10^6. This represents about 20% of MSF unit cost with its auxiliaries [2]. So, the MSF unit cost can be estimated by US$ 38.64 × 10^6 (or $ 6.44/gallon). The water Desalination Report of 1989 [3] showed significant increase in the cost of Cu-Ni tubing from the end of 1988 to mid 1989 [Cu-Ni (90:10) cost jumped form US$ 3500/ton to US$ 7407/ton while Cu-Ni (70:30) cost jumped from US$ 4200/ton to US$ 5/gallon to at least US$ 7.5/gallon in 1989. This cost does not include the costs of any civil, intake, or any auxiliary work. According to Ericsson et al. [5], the MSF unit cost represents about 46% of the total plant cost, i.e. the total plant capital cost is $ 84 × 10^6 (or US$ 14/gallon).

It may be interesting to notice that if the used $1\frac{3}{4}$ in. diameter tubes were put side by side, their length would reach 652 km (18.3 m tube length × 35,650 tube number). The solid volume of the tube material (tube thickness of 1.22 mm) is 113 m^3 (equivalent to 1000 tons). The weight of a typical 6-migd evaporator is about 2500 tons when empty, 8000 tons when it is filled with water. The installation of such an evaporator requires substantial supporting civil work for foundation.

Shell

A rectangular vessel (or shell) is usually used to provide a series of stages. Each stage (see Figure 2) should have: (i) an area for the brine to flash; (ii) large space for large brine droplets to separate from the flashed vapor; (iii) enough volume to house the condenser tube bundle with its distillate tray and venting means, components; (iv) area to house the demisters; and (v) gates or openings to control the flashing brine stream flow from one stage to another and the accumulated distillate respectively. The vessel should be rigid enough to withstand the pressure (either positive or negative), and have allowance for material loss by corrosion. The vessel material is cladded Carbon Steel (say of 6 mm thickness). Cladding of the Makkah-Taif plant [1] are stainless-steels 317 for bottom, 316 L for side walls, and 340 L for tops. Stage dividers are made of stainless-steel 316.

The tube sheets are made of the same material of the tubes while the water boxes are usually made of carbon steel lined Cu-Ni (90:10). For our design of 6 migd, the average stage dimensions are 3.3 m length. 17.6 m width and 4.2 m height. So, a single tier of 24 stage has a length of 80 m, 24 divider, volume of 5910 m^3, and requires an area of 1800 m^2 (if 2 m is left at each side).

MSF Pumps

Several pumps are required to operate the MSF unit. These include seawater supply pump, brine recirculation pump, distillate pump, brine blowdown pump, heater drain pump, seawater recirculation pump (to control the temperature of the last stage), chemical injection pumps, etc. Table 1 gives some of the calculated pump flowrates and heads and the nominal power rating of their motors.

Table 1. Pumps used in the MSF unit.

No.	Pump duty	Flow-rate (l/s)	Head (m)	Motor rating (kW)
1.	Recirculating pump	4200	73	4100
2.	Seawater cooling pump	2675	25	890
3.	Distillate pump	348	57	280
4.	Blowdown pump	528	23	170
5.	Condensate pump	45	114	90
6.	Seawater recirculation pump	750	15.8	170

3. DESALINATION PLANT SUSTAINABILITY INDICATORS

The sustainability assessment of desalination plant should include different criteria reflecting important aspects immanent to the water production. In this respect, attention has been focused on resources, environmental, social and economic aspect of the desalination plant. Each of the criteria will be measured by the respective indicators. These indicators have to meet a specific requirements, including:

- to reflect sustainability concept. This will imply that the indicators for respective criteria represent quantities, which have relevancy to the sustainability to be defined with indicators which can be measured and are available as physical parameters representing the data which are possible to obtain with respective quantitative or qualitative form.

- to be based on the reliable information. In this respect the indicators have to be the data which you must trust because their totality may be the milestone for the important decision to be made.
- to give possibility to perform optimisation of the system to minimise energy cost under conditions associated with available material, financial resources, protection of the environment and government regulations, together with safety, reliability, availability and maintainability of the system..

Indicators satisfying mentioned prerequest will be used for the assessment of individual options under consideration.

For desalination plants assessment the following indicators will be used:

3.1 Resource Indicators

Within the group of resource indicators for desalination plants are Fuel Resource Indicator and Material Resource Indicator defined as specified in the Table 2.

Table 2. Fuel Resource Indicator and Material Resource Indicators.

Name	Definition	Dimension
Fuel Resource Indicator	Amount of fuel per m^3 of desalinated water	kg_{fuel}/m^3
Material Resource Indicator	Amount of material used in construction of the plant per m^3 of desalinated water	Kg_{mat}/m^3

3.2 Environment Indicators

Environment indicators comprise reflection of the environment pollutants produced by the desalination plant. These group of indicators strongly depend on the quality of the fuel. In our assessment we will take note of three gases produced by desalination plants, namely: CO_2, SO_2 and NO_x. In this respect three indicators are defined as specified in Table 3.

Table 3. Environment indicators.

Name	Definition	Dimension
CO_2Environment Indicator	Amount of CO_2 per m^3 of desalinated water	kg_{CO2}/m^3
SO_2 Environment Indicator	Amount of SO_2 per m^3 of desalinated water	kg_{SO2}/m^3
NO_x Environment Indicator	Amount of NO_x per m^3 of desalinated water	kg_{NOx}/m^3

3.3 Economic Indicator

Economic indicator is defined by a unit cost of desalinated water. The cost will include capital cost, operation and maintenance cost and fuel cost. Even the date on economic validation o of individual plants are proprietary and scarce, in this evaluation we will use publicly available information. The definition of economic indicator is given in Table 4.

Table 4. Economic indicators.

Name	Definition	Dimension
Economic Indicator	Cost of m^3 of desalinated water	USD/m^3

4. DESALINATION PLANT OPTION SELECTION

4.1 Option 1 - Single Purpose MSF Desalination Plants

A single purpose desalination plant using MSF desalination units requires the use of boilers to supply the desalting units with its thermal energy needs (as steam at moderately low pressure of 1-3 bars), and to supply steam ejectors to eject non-condensable gases from the units. This is a wasteful process from thermodynamics principles, since high fuel of high availability is burned to produce stem of low availability. This can be clearly shown by calculating the second law efficiency of boilers producing saturated steam at atmospheric pressure (about 15%), and that producing steam at conditions required by modern power plants (160 bar, and 550°C) with second law efficiency of 46%. The average thermal energy consumption of a typical MSF used in Kuwait of GR equal to 8, (GR = the kg of desalted water/kg of supplied steam), is 320 MJ/ m^3 including steam supplied to the ejectors. The boiler efficiency for this moderately low-pressure steam, say 80%, is lower than the efficiency of steam generators producing high availability steam (about 88%). Then, a fuel energy consumption to produce 1 m^3 by directly boiler driven MSF units is 400 MJ/m^3. Additional mechanical energy is needed to drive the system pumps, in the range of 4.5 kWh/m^3 (16.2 MJ/m^3 mechanical work). To produce this amount of energy from a power plant of 0.36 efficiency, the fuel energy needed is 45 MJ/m^3. So, the amount of fuel energy required is 445 MJ/m^3 of desalted water. So, the Fuel resource Indicator for desalting water, expressed by the fuel energy consumed to produce 1 m^3 of desalted water, by directly boiler driven MSF system is RI (fuel/water) = 455 (MJ/m^3)/40 (MJ/kg fuel) = 11.225 kg fuel/m^3 desalted water. For this case of directly boiler driven desalting system, other sustainability indicators can be calculated as follows in Table 5.

Table 5. Single Purpose MSF Desalination Plants Indicators.

RI_{fuel}	EI_{CO2}	EI_{SO2}	EI_{NOx}	EcI
kg_{fuel}/m^3	kg_{CO2}/m^3	kg_{fSO2}/m^3	kg_{NOx}/m^3	USD/m^3
11.225	37	0.0898	0.0578	2.66

The average daily desalted water production from MSF desalting units in Kuwait in 1997 was 187.7 MGD or (187.7 *4550 m^3/Million Gallons) = 854.035 thousands m^3/day. So, the annual desalted water production in 1997 was 311.723 million of m^3. The sustainability indicators for the case of direct boiler driven desalting units producing 311.723 million of m^3/year are: RI (fuel/water) = 11.225 kg fuel/ m^3 desalted water * 311.723 million m^3/year = 3.499 million tons of fuel/year. The annual fuel cost for water production by single purpose desalting system = 3.499*175.81 = 615.159 million dollars/year The specific investment cost of large MSF desalting plants, according to United Nations Repot, DEEP 1998, is \$2153/(m^3/day). In Kuwait, the installed capacity of the desalting plants is 234 MGD (1.0647 million m^3/day), while the average production in 1997 is 178.7 MGD (0.813 millions m^3/day), i.e. the operating load factor is 0.802. The installed capacity of desalting units is usually planned to satisfy the desalted water needs for a long time (usually 10 years) with the low load factor in the early years of the plant operation. Then the average load factor over the time span of the desalting plants is much lower than 0.802, say 0.65. Then, for expected plant life of 25 years, the specific capital cost capital cost is \$2153/(25* 0.65*365) = \$0.363/m^3. The annual capital cost distributed over the plant life is 1.0647 *0.65* 365* 0.363 = \$91.694 millions per year. If the operating and maintenance cost represent 1.2 of the fuel cost, then the annual operating cost of desalted water is 615.159*1.2 = 738.191 Then the total annual cost of producing 311.723 million m^3 in a year is equal to 738.191 + 91.694 = 829.885 million dollars, and this gives a specific cost = 829.885/311.723 = \$2.66225/m^3.

4.2 Option 2 - Dual Purpose Power-MSF Desalting Plant

When the plant is working as dual purpose plant and producing 225 MWe power and 12 MGD water, the fuel energy input to steam generator, according to the design conditions, is increased to 702.6 MWth, compared to fuel energy input of 610 MWth when only 225 MWe electric was produced. So, an additional 92.6 MWth fuel energy is added at the steam generator in order to supply 180 MWth to the desalting units. The cost of fuel supply to the steam generator has to be charged to both products, power and desalted water, according to an accepted rule. If all the benefit of dual-purpose plant is given to the desalting process, then 92.6 MWth is being charged to the desalting process for its thermal energy input. However this is not fair and both production should be benefited for the combination of the power and desalting units arrangement. The steam extracted from the turbine to the desalting units would produce additional 42.82 MWe of work when expanded in the turbine to the condensing conditions, instead of being extracted to the desalting units. This equal to 3.57 Mwe equivalent mechanical work per one million gallons per day, or 67.1 kJ/kg mechanical work per unit distillate, to account for the thermal energy input (i.e.18.84kWh/m^3). To produce this amount of mechanical energy in a power plant of 0.369 efficiency (design conditions), 116.043 MWth fuel energy is needed to produce 12 MGD, i.e.183.629 kJ/kg. So, the ratio of the fuel to be charged to the desalting units should be 42.82/(42.82 + 225) = 0.16. This means that 116.1 MWth has to be charged for the thermal energy to desalt 12 MGD. Meanwhile the balance, 586.5 MW is to be charged to the power production of 225 MW. This shows a significance decrease in fuel energy charged to desalting process, 116.1 compared to 180 MW actually added to the desalting units. As for the power production slight decrease in fuel energy is charged, 586.5 MW, instead of 610 MW for single purpose power plant. This is equivalent to raising the design efficiency from 0.369 (for single purpose power plant) to 0.3836, or decreasing the design heat rate to 9384 kJ/kWh (compared to 9756 kJ/kWh for single purpose power plant). All these numbers are for design conditions. It is noticed here that additional pumping energy should be considered (16.2 kJ/kg) for the desalting process. Then the total mechanical work accounting for thermal energy supplied to the brine heater and pumping energy is 83.3 kJ/kg mechanical work. Again to produce this amount of work, the fuel energy required is 83.3/0.369 = 225.75 kJ/kg, (225.75 MJ/m^3).

It is noticed here that the fuel energy consumed for desalting seawater is 225.75 MJ/m^3 (compared to 455 MJ/m^3 in a single purpose desalting plant). This means a 50.4% decrease in fuel energy when dual-purpose power water plant is used compared to single purpose desalting plant. So, the following sustainability factors for desalting water production will decrease by the same ratio to become:

The cost of desalted water can be calculated by using the same assumptions made earlier. The specific capital cost capital cost is \$2153/(25* 0.65*365) = \$0.363/m^3. The annual capital cost distributed over the plant life is \$252.6 millions per year. If the operating and maintenance cost represent 1.2 of the fuel cost, then the annual operating cost of desalted water is 305.144 * 1.2 = 366.1728. The total annual cost of producing 311.723 million m^3 by dual-purpose plant per year is equal to 366.1728 + 91.694 = 457.8668 million dollars, and this gives the specific cost = 457.694/311.723 = \$ 1.469/m^3. In table 6 are shown data for Dual Purpose Power-MSF Plant.

Table 6. Dual Purpose Power-MSF Desalting Plant

RI_{fuel}	EI_{CO2}	EI_{SO2}	EI_{NOx}	EcI
kg_{fuel}/m^3	kg_{CO2}/m^3	kg_{fSO2}/m^3	kg_{NOx}/m^3	USD/m^3
6.967	21.16	0.0556	0.0359	1.469

4.3 Option 3 - Reverse Osmosis Desalting Plants with Electric Energy from Main Line

Reverse Osmosis desalting system is the main competitor of the MSF desalting system. The RO system became more attractive by the continuous improvements in membrane materials, the raising of both feed pressure and temperature limits, and production of potable water from high salinity water in the Gulf area.

Reverse Osmosis desalting plant is used, with maximum guaranteed energy consumption of 7.5 kWh/m^3 product water (27 MJ mechanical work/m^3 product, or 77.14 MW fuel energy/m^3), the annual fuel energy consumption would be 643.07 millions tons per year.

Our concern here is to compare the energy consumption of the RO system and compares it with that of the MSF system. Consider a practical case similar to Jeddah 1 RO plant phase II of 12.5 MGD reported by Al-Badawi et al. (1995). The plant consists of 10 trains. Each train gives a product rate of 5680 m^3/day (65.74 kg/s).The procedure to calculate the power consumption is outlined here for one train of this Jeddah plant.

Since the conversion ratio (product/feed) is 0.35, then the feed flow rate is 187.83 kg/s. The plant actual feed pressure is around 60 bar (the membranes maximum allowable pressure is 70 bars. Assuming an efficiency of the feed pump equal to 0.75, and its driving motor is 0.92, then the feed pump power consumption is Qf (in m^3/s) * P (in kPa)/(Ep* Em) = (187.83/1000) * (6000)/(0.75*0.92) = 1633.3 kW.

By considering 20% more energy is consumed by other pumps (e.g. seawater supply, seawater boost, and chemical dosing pumps) the power consumption is 1.2 * 1633.3 = 1959.96 kW To calculate the energy recovered in a turbine from the brine blow-down: the brine flow rate = 187.83 - 65.74 = 122.09 kW The brine pressure = feed pressure - pressure loss in the feed-brine side = 60 - 3 = 57 bars Recovered energy = brine flow rate (in m^3/s) * P (in kPa) * Et = (122.09/1000) * 5600 * 0.65 = 444.41 kW Net energy consumption = 1959.96 - 444.41 = 1515.55 kW Specific work done = 1515.55/65.74 = 23.054 kJ/kg = 6.404 kWh/m^3 It may be noticed here that AlFujaira plant actual measured power consumption is 6.54 kWh/m^3 (Schierah 1955), but the consumption is 7.5 kWh/m^3.

This plant is in operation since 1991. Now to calculate the sustainability indexes when RO are used, the fuel energy required to produce 7.5 kWh/m^3 (27 MJ/m^3) is equal to 27/0.369 = 73.17MJ/m^3. So, the mass of fuel required to produce 1 m^3 of desalted water is 73.15/40 = 1.83 kg of fuel/kg of water.

Again the specific investment cost of large Reverse Osmosis desalting plants, according to United Nations Repot, DEEP 1998, is $1280/$m^3$/day. By assuming the same average load factor considered before over the time span of the desalting plant, say 0.65. Then, for expected plant life of 25 years, the specific capital cost capital cost is $1280/(25* 0.65*365) = $0.216/$m^3$. The annual capital cost distributed over the plant life is 1.0647 *0.65* 365* 0.216 = $54.562 millions per year. If the operating and maintenance cost represent 1.3 of the consumed electric energy cost, then the annual operating cost of desalted water, by assuming that the cost of 1 kWh is 20 Kuwaiti fils (about $0.065/kWh). Then the operating annual cost of producing 311.723 million m^3 in a year is equal to 311.723*(7.5 kWh/m^3)*0.065 = 151.965 million dollars/year. Then the annual total cost of producing 311.723 million m^3 by RO system is 151.965 + 54.562 = 206.527 million dollars/year This gives a unit cost of $0.6625/$m^3$ So, the mass of fuel required to produce 1 m^3 of desalted water is 73.15/40 = 1.83 kg of fuel/kg of water.

Then RI (fuel/water) = 1.8323 kg fuel/m^3 desalted water; EI (CO2) = 6.0403 kg/m^3;EI (SO2) = 0.004825 kg/m^3;EI (NOx) = 0.009436 kg/m^3 are presented in Table 7.

Table 7. Reverse Osmosis Desalting Plants with Electric Energy from Main Line.

RI_{fuel}	EI_{CO2}	EI_{SO2}	EI_{NOx}	EcI
kg_{fuel}/m^3	kg_{CO2}/m^3	kg_{fSO2}/m^3	kg_{NOx}/m^3	USD/m^3
1.832	6.04	00048	0.0094	0.6625

4.4 Option 4 - Reverse Osmoses Desalination Plant with PV Electric Energy Production

This option will take into a consideration evaluation of the desalination plant with the same parameters as in option 3 but energy supply will be from the PV solar power plant. This will imply that the Fuel Resource Indicators and Environment Indicators will be same as for Option 3 but Economic Indicator will be changed due to change in electric energy cost. The specific energy consumption for the desalination plant with reverse osmoses is 7.5 kWh/m^3 so that for the plant under consideration the total electric energy consumption is 1633.3 kW. Average irradiation 6.1 kWh/m^2 d or 150 kWh/m^2 a which will give production 1600 kWh$_{AC}$/d. It is assumed 0.30 USD/kWh so the total cost of photovoltaic electric energy. With the desalination water production 312 x 10^6 m^3 the total operational cost is 700x10^6 USD so that with annual capital cost for desalination plant equal to 54.6 USD/m^3 the total annual cost is 755.4x10^{6}USD or specific cost per unit desalinated water 2.42 USD/m^3. Since there is no fuel consumption the Resource Indicator and Environment Indicators are equal zero.

Table 8. Reverse Osmoses Desalination Plant with PV Electric Energy Production.

RI_{fuel}	EI_{CO2}	EI_{SO2}	EI_{NOx}	EcI
kg_{fuel}/m^3	kg_{CO2}/m^3	kg_{fSO2}/m^3	kg_{NOx}/m^3	USD/m^3
0	0	0	0	2.42

5. EVALUATION OF THE SUSTAINABILITY INDICATORS

Next step in the assessment of desalination plants is to determine sustainability indicators. In this respect, in this analysis the following indicators are taken into consideration: Fuel Resource Indicators, CO_2 Environment Indicator and Cost Economic Indicator, shown in Table 9.

Table 9. Fuel Resource Indicators, CO_2 Environment Indicator and Cost Economic Indicator.

Option		RI	EI	EcI
		kg/m^3	kg/m^3	kg/m^3
Option	1	11,225	37	2,66
Option	2	6,967	21,16	1,47
Option	3	1,883	6,04	0,66
Option	4	0	0	2,82
Max		11,225	37	2,82
Min		0	0	0,66
Max-Min		11,225	37	2,16

5.1 The Sustainability Assessment Procedure

The sustainability assessment procedure is based on the Decision Support System's Shell (DSSS) ASPID-3W [5,6], which is a computer realisation of so called ASPID-method (ASPID [7].

It is assumed that a set $X=\{x^{(j)},\ j=1,...,k\}$ of k *options* is fixed. In our cases are four *options* of the desalination plant ($k=4$): $x^{(1)}$ - option 1 (Single Purpose MSF Desalination Plan); $x^{(2)}$ - option 2 (Dual Purpose Power –MSF Desalination Plant); x^{3} - option 3 (Reverse Osmoses Desalination Plant with local electric energy source); $x^{(4)}$ - option 4 (Reverse Osmoses Desalination Plant with PV electric energy source).

Each option $x^{(j)}$ is identified with a vector $x^{(j)}=(x_1^{(j)},...,x_m^{(j)})$. A component $x_i^{(j)}$ of the vector $x^{(j)}$ is treated as a value of an initial x_i of the option $x^{(j)}$. In our case initial *attributes* are three *indicators* of the desalination plant sustainability: x_1 - Resource Indicator (RI); x_2 - Environment Indicator (EI); x_3 - Efficiency Indicator (FI).

So, the vector $x^{(j)}=(x_1^{(j)},...,x_m^{(j)})$ may be treated as a value of the attribute-vector $x=(x_1,...,x_m)$. It is supposed that each indicator x_i is necessary and the whole attribute-vector x is sufficient for a fixed *quality* of option definition. In the context of decision making the quality under consideration may be identified with the *preference* of the option for a decision-maker.

A fixed quality of option $x^{(j)}$, $j=1,...,k$, is defined by *criteria* $q_1,...,q_m$ each of them being a function of a corresponding indicator: $q_i=q_i(x_i)$, $i=1,...,m$. A value $q_i^{(j)}=q_i(x_i^{(j)})$ of a quality level (degree of preference) of j-th option from the point of view of i-th *specific criterion*. Without the loss in generality it may be supposed that all specific criteria are normalised, i.e., any criterion q_i meets the inequality $0\le q_i\le 1$. As this *normalisation* takes place, the minimal value $q_i^{(j)}=0$ of i-th criterion is correlated with an object $x^{(j)}$, which has minimal degree of preference (from the point of view of i-th criterion), and maximal value $q_i^{(l)}=1$ - with an option $x^{(l)}$ which has maximal degree of preference (from the same point of view). So, every option $x^{(j)} \lfloor X$ gets a *multicriteria indicator* $q^{(j)}=(q_1^{(j)},...,q_m^{(j)})$, $0\le q_i^{(j)}\le 1$, which may be treated as a value of the criteria-vector $q=(q_1,...,q_m)$.

Let us consider linear normalisation function $q_i=q_i(x_i)$. If preference of any option is increasing (from the point of view of i-th criterion) with increasing of the argument x_i then we shall use increasing normalisation function determined by the formula

$$q_i(x_i;\theta)=\begin{cases} 0 & \text{if} \quad x_i \le MIN_i, \\ \dfrac{x_i - MIN_i}{MAX_i - MIN_i} & \text{if} \quad MIN_i < x_i \le MAX_i, \\ 1 & \text{if} \quad x_i > MAX_i. \end{cases} \tag{1}$$

If preference of any option is decreasing (from the point of view of i-th criterion) with increasing of the argument x_i then we shall use decreasing normalisation function determined by the formula

$$q_i(x_i;\theta)=\begin{cases} 1 & \text{if} \quad x_i \le MIN_i, \\ \dfrac{MAX_i - x_i}{MAX_i - MIN_i} & \text{if} \quad MIN_i < x_i \le MAX_i, \\ 0 & \text{if} \quad x_i > MAX_i. \end{cases} \tag{2}$$

The value MIN_i (MAX_i) in (2) may be chosen by the simplest way as the minimal (maximal) value $MIN_i = \min_j\{x_i^{(j)}, j=1,...,k\}$ ($MAX_i = \max_j\{x_i^{(j)}, j=1,...,k\}$) from the set $\{x_i^{(j)}, j=1,...,k\}$.

This linear normalisation gives in our case values of normalised indicator: $q_1 = q_1(x_1) = q_1(RI)$, $q_2 = q_2(x_2) = q_2(EI)$, $q_3 = q_3(x_3) = q_3(EcI)$,. So, j-th row in Table 9 may be treated as a multicriteria indicator $q^{(j)} = (q_1^{(j)},...,q_4^{(j)})$ of j-th option, $j=1,...,4$.

Table 10. Values of normalised Sustainability Indicators of the Desalination plant options.

		q-1	q-2	q-3
Option	1	0	0	0
Option	2	0,379332	0,428108	0,5918
Option	3	0,832249	0,836757	1
Option	4	1	1	0.1327

The values $q_i^{(j)}$, $i=1,\ldots,m$, $j=1,...,k$, of specific indices $q_1,\ldots,q_m$ for all options from the set X being fixed we can try to compare options *general preference* with help of component-wise preference relation $\succ$ determined on the $X = \{x^{(j)}, j = 1,\ldots,k\}$ by the condition

$$\forall\, x^{(j)}, x^{(l)} \llcorner X\ (x^{(j)} \succ x^{(l)}) \ulcorner\ ((\forall i\ q_i^{(j)}\ ?\ q_i^{(l)})\ \&\ (\exists s\colon q_s^{(J)} > q_s^{(l)})). \quad (3)$$

In plain words, option $x^{(j)}$ is more preferable "in general" than option $x^{(l)}$ ($x^{(j)} \succ x^{(l)}$) if and only if option $x^{(l)}$ is not more preferable than $x^{(j)}$ from the point of view of each specific criterion q_i ($q_i^{(j)}\ ?\ q_i^{(l)}, i = 1,\ldots,m$) and there exists a specific criterion q_s such that $x^{(j)}$ is more preferable than $x^{(l)}$ from the point of view of the criterion ($q_s^{(j)} > q_s^{(l)}$).

Two options $x^{(j)}, x^{(l)} \llcorner X$ are named *incomparable* if they meet the condition

$$(\exists r\colon q_r^{(j)} > q_r^{(l)})\ \&\ (\exists s\colon q_s^{(j)} < q_s^{(l)}). \quad (4)$$

For example, multicriteria indicator $q^{(1)} = (q_1^{(1)},\ldots,q_3^{(1)}) \cup ()$, $q^{(2)} = (q_1^{(2)},\ldots,q_3^{(2)}) \cup ()$ of options 1,2 sustainability are incomparable as Option 2 is "better" (more preferable) than Option 1 by the criteria q_1, q_2, q_3, but Option 1 is "better" than Option 2 by the criterion q_4.

The problem of options multicriteria incomparability may be decided by synthesis of specific criteria $q_1,\ldots,q_m$ into one *general criterion* (index) Q determined by a scalar function $Q = Q(q) = Q(q_1,\ldots,q_m)$ which meets the condition of monotony $((x^{(j)} \succ x^{(l)})\ ?\ (Q(q^{(j)})\ ?\ Q(q^{(l)})))$ and fills the requirements: (1) $Q(0,\ldots,0) = 0$; (2) $Q(1,\ldots,1) = 1$. Inequality $Q(q^{(j)}) > Q(q^{(l)})$ means that j-th option is more preferable than l-th object. Now all options are comparable as there are only three alternatives for any pair of objects $x^{(j)}$, $x^{(l)}$: $Q(q^{(j)}) > Q(q^{(l)})$; $Q(q^{(j)}) < Q(q^{(l)})$; $Q(q^{(j)}) = Q(q^{(l)})$.

The synthesising functions Q is the generalised weighted mean, namely *additive function* (*weighted arithmetical mean*)

$$Q_+(q;w) = \overset{m}{\underset{i=1}{}}\ w_i q_i\,. \quad (5)$$

In this formula $w = (w_1,\dots,w_m)$, w_i ? 0, $w_1+\dots+w_m = 1$, is a vector of *weight-coefficients* $w_1,\dots,w_m$ (*weight-vector*). A weight-coefficient w_i is a *measure of relative significance* of the corresponding specific criterion q_i for aggregate estimation $Q(q^{(j)})$ of general preference of an option $x^{(j)}$ ⌊ X.

6. DISCUSSION OF THE GENERAL INDEX OF SUSTAINABILITY

The analysis is based on the results obtained by the DSSS procedure. In this analysis we will take into a consideration different alternative as regard weighting factors. In our analysis of the effect of individual criteria on the General Index of Sustainability it is assumed three different cases which will reflect changes in the mutual relation of the weighting factors on the decision making process.

Following cases are taken into a consideration:

1. $I = I_1 = w_1 = w_2 > w_3$ - there is no information about admissible weight-coefficients in our disposal
2. $I = I_2 = \{w_1 > w_2 > w_3\}$ - specific criteria are strictly ranked by their influence on General Sustainability Index.
3. $I = I_3 = \{w_3 > w_2 > w_1\}$ - we have the ordinal information I_3 about weight-coefficients.

For all three cases input data for DSSS are the same:
Number of option k = 4.
Number of sustainability indicators m = 3.
Number of steps n = 20. So, it is supposed that the measurement of weight-coefficients is accurate to within the step h = 1/n = 1/20 = 0.05.

Number of all possible variants (of all weight-vectors $w = (w_1,\dots,w_m)$)
N(3;20) = 221

6.1 Case 1

The Case 1 comprise situation when all weight-coefficients are equal

$w_1 = w_2 > w_3$

Even we know that it is not realistic situation, it gives possibility to make assessment of the contribution of individual indicators to the General Index for the priority definition. In table 11 and table 12 are presented data for General Index and probability of pair-wise dominancy for Case 1.

Table 11. General Index for Case 1.

	Opt.1 MSF (single)	Opt2 MSF (Dual)	Opt.3 Loc. Elect.	Opt.4 PV elect.
General index Q (q,w)	0.0000	0.4648	0.8914	0.7109
Dis Q	0.0000	0.0000	0.0000	0.0000

Table 12. Probability of pair-wise dominancy for Case 1.

	Opt.1	Opt.2	Opt.3	Opt.4
Opt.1 – MSF (single)	0.0000	0.0000	0.0000	0.0000
Opt.2 – MSF (dual)	1.0000	0.0000	0.0000	0.0000
Opt.3 – Loc. Elect.	1.0000	1.0000	0.0000	0.0000
Opt.4 – PV elect.	1.0000	1.0000	0.0000	0.0000

From the Figure 1 it can be noticed that the option 3 – RO Desalination Plant with local electric supply is the first on the priority list based on the General Index of Sustainability.

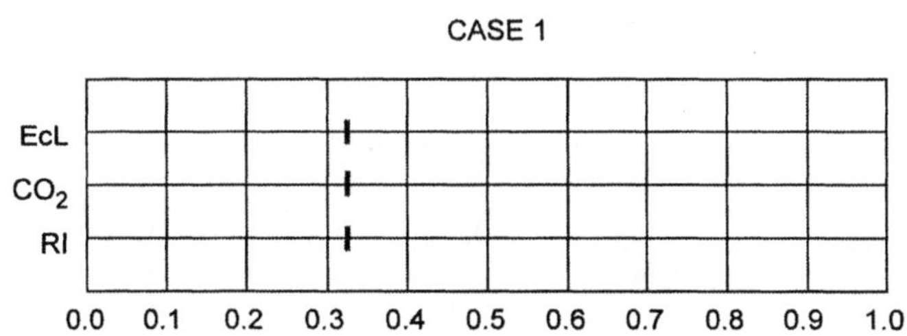

Figure 1. Normalised indicator values for Case 1.

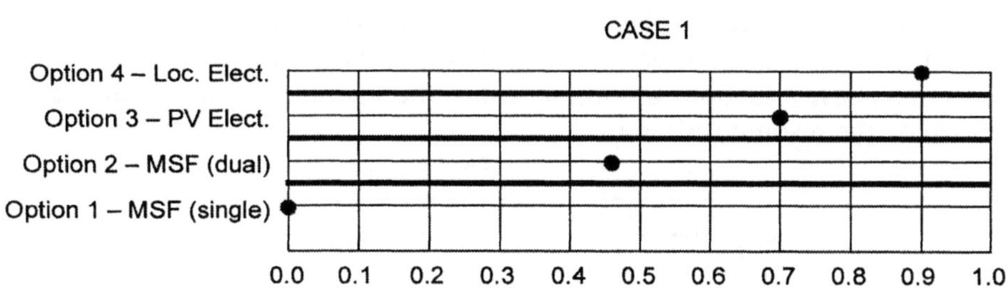

Figure 2. General Index for case 1.

6.2 Case 2

Case 2 represents situations when

$w_1 > w_2 > w_3$

From total number of variants satisfying adapted conditions we will obtain that only N = 29 will meet this requirement. If the General Index of Sustainability will be determined for all those variants the average value of the General Index of Sustainability is obtained with respective standard deviation. Figure 2 shows graphical presentation of the General Index of Sustainability. Under imposed condition of weighting factors mutual relation the priority list based on the General Index of Sustainability is obtained. Option 4 – RO Deasalination Plant with PV electric energy supply is the first on the list. It should be noticed that RO Desalination Plant with local electric energy supply is on the second place followed by the Dual Purpose MSF Desalination Plant. In table 13 and table 14 and Figures 3 and 4 are presented data for General Index and probability of pair-wise dominancy for Case 2.

Table 13. General Index for Case 2.

	Opt.1 MSF (single)	Opt2 MSF (Dual)	Opt.3 Loc. Elect.	Opt.4 PV elect.
General index Q (q,w)	0.0000	0.4111	0.8536	0.9106
Dis Q	0.0000	0.0193	0.0129	0.0682

Table 14. Probability of pair-wise dominancy for Case 2.

	Opt.1	Opt.2	Opt.3	Opt.4
Opt.1 – MSF (single)	0.0000	0.0000	0.0000	0.0000
Opt.2 – MSF (dual)	1.0000	0.0000	0.0000	0.0000
Opt.3 – Loc. Elect.	1.0000	1.0000	0.0000	0.222
Opt.4 – PV elect.	1.0000	1.0000	0.778	0.0000

As it can be noticed the Case 2 reflect those situations when priority is given to the resource criteria in comparison to the environment and economic criteria. These situations are immanent to the variants leading to the decision based on the resource criteria.

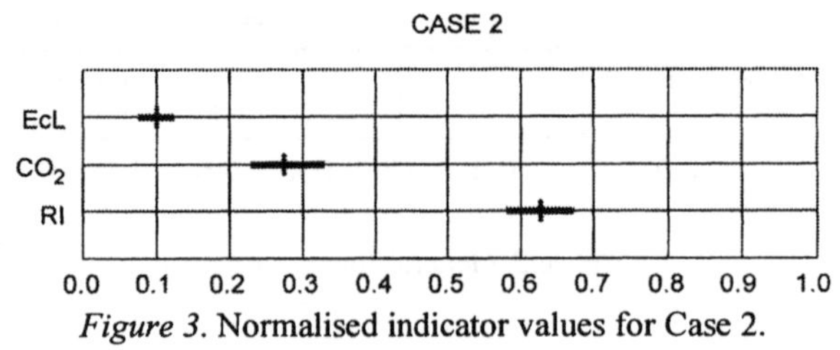

Figure 3. Normalised indicator values for Case 2.

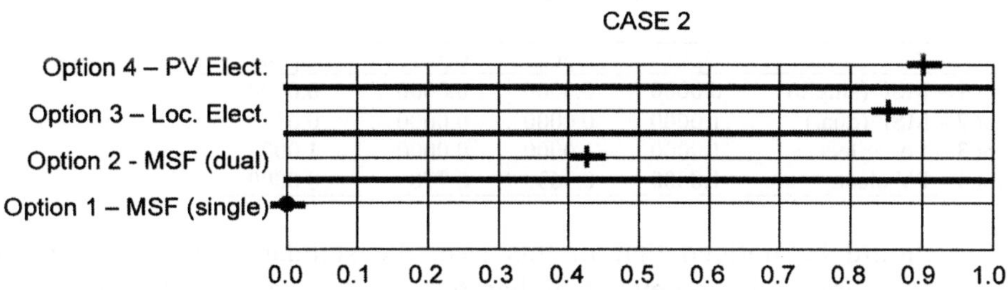

Figure 4. General Index for Case 2.

6.3 Case 3

Case 3 represents the situations when

$w_3 > w_2 > w_1$

Case 3 is designed to emphasise economic criteria in comparison with environment and resource criteria. It should be kept in mind that the adapted procedure reflects also mutual relation of the individual criteria but requires respective priority as specified in this case. Giving priority to economic criteria is of specific interest to investigate how strong are other two criteria and to what extent can contribute to the priority list. It is of interest to notice that the change in the priority list is obtained as the result of the change in the criteria priority. This reflects one of the important characteristics of this method. Change from priority in Case 2 to Case 3 imply the change of the emphasise from the resources priority to economic priority has lead to the change in the priority list of the options under consideration. In table 15 and table 16, and Figures 5 and 6, are presented data for General Index and probability of pair-wise dominancy for Case 3.

Table 15. General Index for Case 3.

	Opt.1 MSF (single)	Opt2 MSF (Dual)	Opt.3 Loc. Elect.	Opt.4 PV elect.
General index Q (q,w)	0.0000	0.5242	0.9381	0.4630
Dis Q	0.0000	0.0263	0.0230	0.1228

Table 16. Probability of pair-wise dominancy for Case 3.

	Opt.1	Opt.2	Opt.3	Opt.4
Opt.1 – MSF (single)	0.0000	0.0000	0.0000	0.0000
Opt.2 – MSF (dual)	1.0000	0.0000	0.0000	0.637
Opt.3 – Loc. Elect.	1.0000	1.0000	0.0000	1.000
Opt.4 – PV elect.	1.0000	0.363	0.778	0.0000

It should be noticed that in this case the standard deviation for the individual option is somewhat higher but not to the extend that it may influence priority list. It is also of interest to mark change in the priority list as regard others options under consideration in this case. Namely, Option 3 – RO Desalination Plant with local electric energy supply has got priority in relation to Option 4 and Option 1. Again, the decision making procedure is showing sensitivity to the change of priority of criteria.

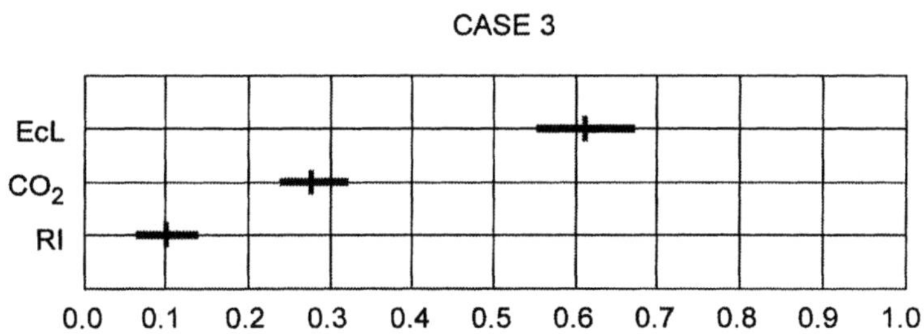

Figure 5. Normalised indicators value for case 3.

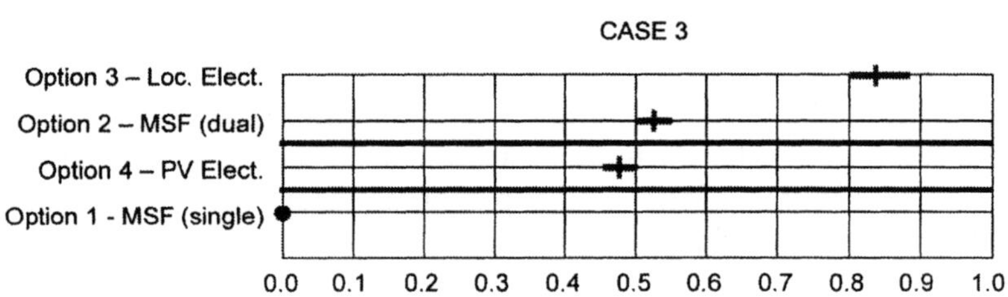

Figure 6. General Index for Case 3.

7. DISCUSSION OF THE RESULTS

Even we should recognise the deficiency of the data used in this analysis, due to its limited accuracy and reliability, it can be seen that the presented examples shows that the potential direction which have to be followed in the future assessment of quality of the desalination plants. It should be emphasised that the demonstrated cases of the desalination plant evaluation proved that the decision making procedure is very much dependent on the priority given to the specific criteria. Also, it is of interest to recognise that the priority list for the selection of appropriate choice is dependent on standard deviation of the specific option. As can be seen there is no difference in standard deviation for the Case 1 but in Case 2 and Case 3 there is a difference in standard deviation for the individual options under consideration.

It should be mentioned that this procedure is exercised by the limited number of possible variants. In the case if the weight-coefficient scale will be adopted with higher accuracy the obtained results will lead to the higher accuracy in the decision making procedure. The accuracy of presented method for sustainability assessment of the desalination plants strongly depend on the accuracy of data used in this type of evaluation. In this analysis, we have used only publicly available data, which has limited quality of decision obtained by the presented method. If this type of the analysis will be done by the respective professional selection of date the reliability of decision could be justified with lower uncertainty.

It is of interest for the decision making process to use more criteria then presented in this analysis. In particular, if the local condition have to be recognised, it may be of interest to present criteria reflecting social indicator and scarcity of material indicator in the design of the plant.

As regard to the selection of desalination plant for the specific location, it will be of interest to take into a consideration not only technical and technological aspect of the water production but also geographical and water scarcity criteria. This may lead to the recognition of the number of options which are of interest for specific conditions.

8. REFERENCES

1. Afgan, N.H., Al Gobaisi, D., Carvalho, M.G., Cumo, M., Sustainable Energy Development, Renewable and Sustainable energy Reviews, 2, pp. 235-286, 1998.
2. Darwish, M., Desalination Comparison, Proceeding of IDA World Congress on Desalination and Water Sciences, Abu Dhabi, Vol. 1, pp. 149-173, 1995.
3. Doroszlai, P.G., Oukili, B., Bouquecha, S., Sea-water Desalination System Based on the Multiple Desalination and Flashing Process, Proceeding of IDA World Congress on Desalination and Water Sciences, Madrid, Vol. 1, pp. 199-225, 1997.
4. Winter, C.J., Sizman, R.L., Vant-Hull L.L., Springer –Verlag, Berlin, 1991.
5. Hovanov, K.N., Hovanov, N.V., DSSS "ASPID–3W". Decision Support System's Shell "Analysis and Synthesis of Parameters under Information Deficiency - 3-d modification for Windows". "ASPID-3W" registered by Federal Agency for Computer Programs Copyright Protection (Russia Federation), Num. 960087, 22 September 1996.
6. Hovanov, N., Kornikov, V., Seregin, I. Qualitative information processing in DSSS ASPID-3W for complex objects estimation under uncertainty//Proceed. Int. Con. "Informatics and Control". St. Petersburg (Russia), pp. 808-816, 1997.
7. Hovanov, N., Analysis and Synthesis of Parameters under Information Deficiency, St. Petersburg. St. Petersburg State University Press., (Monograph, in Russian), 1996.

APPENDIX

UNITED NATIONS CONFERENCE ON ENVIRONMENT & DEVELOPMENT

Conference des Nations Unies sur l'Environnement et le Developpement

160 Route de Florissant
P.0. Box: 80
CH-1231 Conches,
Switzerland
Tel: (41-22)789-1676
Fax: (41-22)789-3536
InterNet/EcoNet: unced@igc.org

AGENDA 21, CHAPTER 35

SCIENCE FOR SUSTAINABLE DEVELOPMENT

NOTE: This is a final, advanced version of a chapter of Agenda 21, as adopted by the Plenary in Rio de Janeiro, on .June 14, 1992. This document will be further edited, translated into the official languages, and published by the United Nations for the General Assembly this autumn.

INTRODUCTION

35.1. This chapter focuses on the role and the use of the sciences in supporting the prudent management of the environment and development for the daily survival and future development of humanity. The programme areas proposed herein are intended to be over-arching, in order to support the specific scientific requirements identified in the other Agenda 21 chapters. One role of the sciences should be to provide information to better enable formulation and selection of environment and development policies in the decision-making process. In order to fulfil this requirement, it will be essential to enhance scientific understanding, improve long-term scientific assessments, strengthen scientific capacities in all countries and ensure that the sciences are responsive to emerging needs.

35.2. Scientists are improving their understanding in areas such as climatic change, growth in rates of resource consumption, demographic trends, and environmental degradation. Changes in those and other areas need to be taken into account in working out long-term strategies for development. A first step towards improving the scientific basis for these strategies is a better understanding of land, oceans, atmosphere and their interlocking water, nutrient and biogeochemical cycles and energy flows which all form part of the Earth system. This is essential if a more accurate estimate is to be provided of the carrying capacity of the planet Earth and of its resilience under the many stresses placed upon it by human activities. The sciences can provide this understanding through increased research into the underlying ecological processes and through the application of modern, effective and efficient tools that are now available, such as remote-sensing devices, robotic monitoring instruments and computing and modelling capabilities. The sciences are playing an important role in linking the fundamental significance of the Earth system as life support to appropriate strategies for development which build on its continued functioning. The sciences should continue to play an increasing role in providing for an improvement in the efficiency of resource utilization and in finding new development practices, resources, and alternatives. There is a need for the sciences constantly to reassess and promote less intensive trends in resource utilization, including less intensive utilization of energy in industry, agriculture, and transportation. Thus, the sciences are increasingly being understood as an essential component in the search for feasible pathways towards sustainable development.

35.3. Scientific knowledge should be applied to articulate and support the goals of sustainable development, through scientific assessments of current conditions and future prospects for the Earth system. Such assessments, based on existing and emerging innovations within the sciences, should be used in the decision-making process and in the interactive processes between the sciences and policy-making. There needs to be an increased output from the sciences in order to enhance understanding and facilitate interaction between science and society. An increase in the scientific capacity and capability to achieve these goals will also be required, particularly in developing countries. Of crucial importance is the need for scientists in developing countries to participate fully in international scientific research programmes dealing with the global problems of environment and development so as to allow all countries to participate on equal footing in negotiations on global environmental and developmental issues. In the face of threats of irreversible environmental damage, lack of full scientific understanding should not be an excuse for postponing actions which are justified in their own right. The precautionary approach could provide a basis for policies relating to complex systems that are not yet fully understood and whose consequences of disturbances cannot yet be predicted.

35.4. The programme areas, which are in harmony with the conclusions and recommendations of the International Conference on an Agenda of Science for Environment and Development into the 21st Century (ASCEND/21) are:

(a) Strengthening the scientific basis for sustainable management;
(b) Enhancing scientific understanding;
(c) Improving long-term scientific assessment;
(d) Building up scientific capacity and capability.

PROGRAMME AREAS

A. Strengthening the scientific basis for sustainable management

Basis for action

35.5. Sustainable development requires taking longer-term perspectives, integrating local and regional effects of global change into the development process, and using the best scientific and traditional knowledge available. The development process should be constantly re-evaluated, in light of the findings of scientific research, to ensure that resource utilization has reduced impacts on the Earth system. Even so, the future is uncertain, and there will be surprises. Good environmental and developmental management policies must therefore be scientifically robust, seeking to keep open a range of options to ensure flexibility of response. The precautionary approach is important. Often, there is a communication gap among scientists, policy makers, and the public at large, whose interests are articulated by both governmental and non-governmental organizations. Better communication is required among scientists, decision makers, and the general public.

Objectives

35.6. The primary objective is for each country with the support of international organizations, as requested, to identify the state of its scientific knowledge and its research needs and priorities in order to achieve, as soon as possible, substantial improvements in:

(a) Large-scale widening of the scientific base and strengthening of scientific and research capacities and capabilities/- in particular, those of developing countries/- in areas relevant to environment and development;
(b) Environmental and developmental policy formulation, building upon the best scientific knowledge and assessments, and taking into account the need to enhance international cooperation and the relative uncertainties of the various processes and options involved;
(c) The interaction between the sciences and decision-making, using the precautionary approach, where appropriate, to change the existing patterns of production and consumption and to gain time for reducing uncertainty with respect to the selection of policy options;
(d) The generation and application of knowledge, especially indigenous and local knowledge, to the capacities of different environments and cultures, to achieve sustained levels of development, taking into account interrelations at the national, regional and international levels;
(e) Improving cooperation between scientists by promoting interdisciplinary research programmes and activities;
(f) Participation of people in setting priorities and in decision-making relating to sustainable development.

Activities

35.7. Countries, with the assistance of international organizations, where required, should:

(a) Prepare an inventory of their natural and social science data holdings relevant to the promotion of sustainable development;
(b) Identify their research needs and priorities in the context of international research efforts;
(c) Strengthen and design appropriate institutional mechanisms at the highest appropriate local, national, subregional and regional levels and within the United Nations system for developing a stronger scientific basis for the improvement of environmental and developmental policy formulation consistent with long-term goals of sustainable development. Current research in this area should be broadened to include more involvement of the public in establishing long-term societal goals for formulating the sustainable development scenarios;
(d) Develop, apply and institute the necessary tools for sustainable

development, with regard to:

(i) Quality-of-life indicators covering, for example, health, education, social welfare, state of the environment, and the economy;
(ii) Economic approaches to environmentally sound development and new and improved incentive structures for better resource management;
(iii) Long-term environmental policy formulation, risk management and environmentally sound technology assessment;

(e) Collect, analyse and integrate data on the linkages between the state of ecosystems and the health of human communities in order to improve knowledge of the cost and benefit of different development policies and strategies in relation to health and the environment, particularly in developing countries;
(f) Conduct scientific studies of national and regional pathways to sustainable development, using comparable and complementary methodologies. Such studies, coordinated by an international science effort, should to a large extent involve local expertise and be conducted by multidisciplinary teams from regional networks and/or research centres, as appropriate and according to national capacities and the available resources;
(g) Improve capabilities for determining scientific research priorities at the national, regional and global levels to meet the needs of sustainable development. This is a process that involves scientific judgements regarding short-term and long-term benefits and possible long-term costs and risks. It should be adaptive and responsive to perceived needs and be carried out via transparent, "user-friendly", risk-evaluation methodologies;
(h) Develop methods to link the findings of the established sciences with the indigenous knowledge of different cultures. The methods should be tested using pilot studies. They should be developed at the local level and should concentrate on the links between the traditional knowledge of indigenous groups and corresponding, current "advanced science", with particular focus on disseminating and applying the results to environmental protection and sustainable development.

Means of implementation

(g) Financing and cost evaluation

35.8. The Conference secretariat has estimated the average total annual cost (1993-2000) of implementing the activities of this programme to be about $1 50 million, including about $30 million from the international community on grant or concessional terms. These are indicative and order of magnitude estimates only and have not been reviewed by Governments. Actual costs and financial terms, including any that are non-concessional, will depend upon, inter alia, the specific strategies and programmes Governments decide upon for implementation.

(b) Scientific and technological means

35.9. The scientific and technological means include the following:

(a) Supporting new scientific research programmes, including their socioeconomic and human aspects, at the community, national, subregional, regional and global levels, to complement and encourage synergies between traditional and conventional scientific knowledge and practices and strengthening interdisciplinary research related to environmental degradation and rehabilitation;

(b) Setting up demonstration models of different types (e.g., socioeconomic, environmental conditions) to study methodologies and formulate guidelines;

(c) Supporting research by developing relative-risk evaluation methods to assist policy makers in ranking scientific research priorities.

B. Enhancing scientific understanding

Basis for action

35.10. In order to promote sustainable development, more extensive knowledge is required of the Earth's carrying capacity, including the processes that could either impair or enhance its ability to support life. The global environment is changing more rapidly than at any time in recent centuries; as a result, surprises may be expected, and the next century could see significant environmental changes. At the same time, the human consumption of energy, water and non-renewable resources is increasing, on both a total and a per/capita basis, and shortages may ensue in many parts of the world even if environmental conditions were to remain unchanged. Social processes are subject to multiple variations across time and space, regions and culture. They both affect and are influenced by changing environmental conditions. Human factors are key driving forces in these intricate sets of relationships and exert their influence directly on global change. Therefore, study of the human dimensions of the causes and consequences of environmental change and of more sustainable development paths is essential.

Objectives

35.1 1. One key objective is to improve and increase the fundamental understanding of the linkages between human and natural environmental systems and improve the analytical and predictive tools required to better understand the environmental impacts of development options by:

(a) Carrying out research programmes in order better to understand the carrying capacity of the Earth as conditioned by its natural systems, such as the biogeochemical cycles, the atmosphere/hydrosphere/lithosphere/cryosphere system, thebiosphere and biodiversity, the agro-ecosystem and other terrestrial and aquatic ecosystems;
(b) Developing and applying new analytical and predictive tools in order to assess more accurately the ways in which the Earth's natural systems are being increasingly influenced by human actions, both deliberate and inadvertent, and demographic trends, and the impact and consequences of those actions and trends;
(c) Integrating physical, economic and social sciences in order better to understand the impacts of economic and social behaviour on the environment and of environmental degradation on local and global economies.

Activities

35.12. The following activities should be undertaken:
(a) Support development of an expanded monitoring network to describe cycles (for example, global, biogeochemical and hydrological cycles) and test hypotheses regarding their behaviour, and improve research into the interactions among the various global cycles and their consequences at national, subregional, regional and global levels as guides to tolerance and vulnerability;
(b) Support national, subregional, regional and international observation and research programmes in global atmospheric chemistry and the sources and sinks of greenhouse gases, and ensure that the results are presented in a publicly accessible and understandable form;
(c) Support national, subregional, regional and international research programmes on marine and terrestrial systems, strengthen global terrestrial databases of their components, expand corresponding systems for monitoring their changing states and enhance predictive modelling of the Earth system and its subsystems, including modelling of the functioning of these systems assuming different intensities of human

impact. The research programmes should include the programmes mentioned in other Agenda 21 chapters which support mechanisms for cooperation and coherence of research programmes on global change;

(d) Encourage coordination of satellite missions, the networks, systems and procedures for processing and disseminating their data; and develop the interface with the research users of Earth observation data and with the United Nations EARTHWATCH system;

(e) Develop the capacity for predicting the responses of terrestrial, freshwater, coastal and marine ecosystems and biodiversity to short- and long-term perturbations of the environment, and develop further restoration ecology;

(f) Study the role of biodiversity and the loss of species in the functioning of ecosystems and the global life-support system;

(g) Initiate a global observing system of parameters needed for the rational management of coastal and mountain zones and significantly expand freshwater quantity/quality monitoring systems, particularly in developing countries;

(h) In order to understand the Earth as a system, develop Earth observation systems from space which will provide integrated, continuous and long-term measurements of the interactions of the atmosphere, hydrosphere and lithosphere, and develop a distribution system for data which will facilitate the utilization of data obtained through observation;

(i) Develop and apply systems and technology that automatically collect, record and transmit data and information to data and analysis centres, in order to monitor marine, terrestrial and atmospheric processes and provide advance warning of natural disasters;

(j) Enhance the contribution of the engineering sciences to multidisciplinary research programmes on the Earth system, in particular with regard to increasing emergency preparedness and reducing the negative effects of major natural disasters;

(k) Intensify research to integrate the physical, economic and social sciences to better understand the impacts of economic and social behaviour on the environment and of environmental degradation on local and global economies and, in particular:

   (i) Develop research on human attitudes and behaviour as driving forces central to an understanding of the causes and consequences of environmental change and resource use;

   (ii) Promote research on human, economic and social responses to global change;

(I) Support development of new user-friendly technologies and systems that facilitate the integration of multidisciplinary, physical, chemical, biological and social/human processes which, in turn, provide information and knowledge for decision makers and the general public.

Means of implementation

(a) Financing and cost evaluation

35.13. The Conference secretariat has estimated the average total annual cost (1993-2000) of implementing the activities of this programme to be about $2 billion, including about $1.5 billion from the international community on grant or concessional terms. These are indicative and order of magnitude estimates only and have not been reviewed by Governments. Actual costs and financial terms, including any that are non-concessional, will depend upon, inter alia, the specific strategies and programmes Governments decide upon for implementation.

(b) Scientific and technological means

35.14. The scientific and technological means include the following:

(a) Supporting and using the relevant national research activities of academia, research institutes and governmental and non-governmental organizations, and promoting their active participation in regional and global programmes, particularly in developing countries;
(b) Increasing the use of appropriate enabling systems and technologies, such as supercomputers, space-based observational technology, Earth- and ocean-based observational technologies, data management and database technologies and, in particular, developing and expanding the Global Climate Observing System.

C. Improving long-term scientific assessment

Basis for action

35.15. Meeting scientific research needs in the environment/development field is only the first step in the support that the sciences can provide for the sustainable development process. The knowledge acquired may then be used to provide scientific assessments (audits) of the current status and for a range of possible future conditions. This implies that the biosphere must be maintained in a healthy state and that losses in biodiversity must be slowed down. Although many of the long-term environmental changes that are likely to affect people and the biosphere are global in scale, key changes can often be made at the national and local levels. At the same time, human activities at the local and regional levels often contribute to global threats/- e.g., stratospheric ozone depletion. Thus scientific assessments and projections are required at the global, regional and local levels. Many countries and organizations already prepare reports on the environment and development which review current conditions and indicate future trends. Regional and global assessments could make full use of such reports but should be broader in scope and include the results of detailed studies of future conditions for a range of assumptions about possible future human responses, using the best available models. Such assessments should be designed to map out manageable development pathways within the environmental and socioeconomic carrying capacity of each region. Full use should be made of traditional knowledge of the local environment.

Objectives

35.16. The primary objective is to provide assessments of the current status and trends in major developmental and environmental issues at the national, subregional, regional and global levels on the basis of the best available scientific knowledge in order to develop alternative strategies, including indigenous approaches, for the different scales of time and space required for long-term policy formulation.

Activities

35.17. The following activities should be undertaken:

(a) Coordinate existing data- and statistics-gathering systems relevant to developmental and environmental issues so as to support preparation of longterm scientific assessments/- e.g., data on resource depletion, import/export flows, energy use, health impacts, demographic trends, etc.; apply the data obtained through the activities identified in programme area B to environment/development assessments at the global, regional and local levels; and promote the wide distribution of the assessments in a form that is responsive to public needs and can be widely understood;

(b) Develop a methodology to carry out national and regional audits and a five-year global audit on an integrated basis. The standardized audits should help to refine the pattern and character of development, examining in particular the capacities of global and regional life-supporting systems to meet the needs of human and non-human life forms and identifying areas and resources vulnerable to further degradation. This task would involve the integration of all relevant sciences at the national, regional, and global levels, and would be organized by governmental agencies, non-governmental organizations, universities and research institutions, assisted by international governmental and non-governmental organizations and United Nations bodies, when necessary and as appropriate. These audits should then be made available to the general public.

Means of implementation

Financing and cost evaluation

35.18. The Conference secretariat has estimated the average total annual cost (1 993-2000) of implementing the activities of this programme to be about $35 million, including about $18 million from the international community on grant or concessional terms. These are indicative and order of magnitude estimates only and have not been reviewed by Governments. Actual costs and financial terms, including any that are non-concessional, will depend upon, inter alia, the specific strategies and programmes Governments decide upon for implementation.

35.19. With regard to the existing data requirements under programme area/A, support should be provided for national data collection and warning systems. This would involve setting up database, information and reporting systems, including data assessment and information dessemination in each region.

D. Building up scientific capacity and capability

Basis for action

35.20. In view of the increasing role the sciences have to play in dealing with the issues of environment and development, it is necessary to build up scientific capacity and strengthen such capacity in all countries/-particularly in developing countries/- to enable them to participate fully in the generation and application of the results of scientific research and development concerning sustainable development. There are many ways to build up scientific and technological capacity. Some of the most important of them are the following: education and training in science and technology; assistance to developing countries to improve infrastructures for research and development which could enable scientists to work more productively; development of incentives to encourage research and development; and greater utilization of their results in the productive sectors of the economy. Such capacity-building would also form the basis for improving public awareness and understanding of the sciences. Special emphasis must be put on the need to assist developing countries to strengthen their capacities to study their own resource bases and ecological systems and manage them better in order to meet national, regional and global challenges. Furthermore, in view of the size and complexity of global environmental problems, a need for more specialists in several disciplines has become evident world wide.

Objectives

35.21. The primary objective is to improve the scientific capacities of all countries/- in particular, those of developing countries/- with specific regard to :

(a) Education, training and facilities for local research and development and human resource development in basic scientific disciplines and in environment-related sciences, utilizing where appropriate traditional and local knowledge of sustainability;
(b) A substantial increase by the year 2000 in the number of scientists/particularly women scientists/- in those developing countries where their number is at present insufficient;
(c) Reducing significantly the exodus of scientists from developing countries and encouraging those who have left to return;
(d) Improving access to relevant information for scientists and decision makers, with the aim of improving public awareness and participation in decision-making;
(e) Involvement of scientists in national, regional and global environmental and developmental research programmes, including multidisciplinary research;
(f) Periodic academic update of scientists from developing countries in their respective fields of knowledge.

Activities

35.22. The following activities should be undertaken:

(a) Promote the education and training of scientists, not only in their disciplines but also in their ability to identify, manage and incorporate environmental considerations into research and development projects; ensure that a sound base in natural systems, ecology and resource management is provided; and develop specialists capable of working in interdisciplinary programmes related to environment and development, including the field of applied social sciences;
(b) Strengthen the scientific infrastructure in schools, universities and research institutions/- particularly those in developing countries/- by the provision of adequate scientific equipment and access to current scientific literature, for the purpose of achieving and sustaining a critical

mass of highly qualified scientists in these countries;

(c) Develop and expand national scientific and technological databases, processing data in unified formats and systems, and allowing full and open access to the depository libraries of regional scientific and technological information networks. Promote submission of scientific and technological information and databases to global or regional data centres and network systems;

(d) Develop and expand regional and global scientific and technological information networks which are based on and linked to national scientific and technological databases; collect, process and disseminate information from regional and global scientific programmes; expand activities to reduce information barriers due to language differences. Increase the applications /- particularly in developing countries/- of computer-based retrieval systems in order to cope with the growth of scientific literature;

(e) Develop, strengthen and forge new partnerships among national, regional and global capacities to promote the full and open exchange of scientific and technological data and information and to facilitate technical assistance related to environmentally sound and sustainable development. This should be done through the development of mechanisms for the sharing of basic research, data and information, and the improvement and development of international networks and centres, including regional linking with national scientific databases, for research, training and monitoring. Such mechanisms should be designed so as to enhance professional cooperation among scientists in all countries and to establish strong national and regional alliances between industry and research institutions;

(f) Improve and develop new links between existing networks of natural and social scientists and universities at the international level in order to strengthen national capacities in the formulation of policy options in the field of environment and development;

(g) Compile, analyse and publish information on indigenous environmental and developmental knowledge, and assist the communities that possess such knowledge to benefit from them.

Means of implementation

(a) Financing and cost evaluation

35.23. The Conference secretariat has estimated the average total annual cost (1 993-2000) of implementing the activities of this programme to be about $750 million, including about $470 million from the international community on grant or concessional terms. These are indicative and order of magnitude estimates only and have not been reviewed by Governments.

Actual costs and financial terms, including any that are non-concessional, will depend upon, inter alia, the specific strategies and programmes Governments decide upon for implementation.

(b) Scientific and technological means

35.24. Such means include increasing and strengthening regional multidisciplinary research and training networks and centres making optimal use of existing facilities and associated sustainable development and technology support systems in developing regions. Promote and use the potential of independent initiatives and indigenous innovations and entrepreneurship. The function of such networks and centres could include, for example:

(a) Support and coordination of scientific cooperation among all nations in the region;
(b) Linking with monitoring centres and carrying out assessment of environmental and developmental conditions;
(c) Support and coordination of national studies of pathways towards sustainable development;
(d) Organization of science education and training;
(e) Establishment and maintenance of information, monitoring and assessment systems and databases.

(c) Capacity-building

35.25. Capacity-building includes the following:

(a) Create conditions (e.g., salaries, equipment, libraries) to ensure that the scientists will work effectively in their home countries;
(b) Enhance national, regional and global capacities for carrying out scientific research and applying scientific and technological information to environmentally sound and sustainable development. This includes a need to increase financial resources to global and regional scientific and technological information networks, as may be appropriate, so that they will be able to function effectively and efficiently in satisfying the scientific needs of developing countries. Ensure the capacity-building of women by recruiting more women in research and research training.

INDEX

E

F

G

H

I

J

K

9 780792 378761

MIX
Papier aus verantwortungsvollen Quellen
Paper from responsible sources
FSC® C105338

If you have any concerns about our products,
you can contact us on
ProductSafety@springernature.com

In case Publisher is established outside the EU,
the EU authorized representative is:
Springer Nature Customer Service Center GmbH
Europaplatz 3, 69115 Heidelberg, Germany

Printed by Libri Plureos GmbH
in Hamburg, Germany